THE SHADOW

OF THE TELESCOPE:

A Biography of John Herschel

THE SHADOW

OF THE TELESCOPE:

A Biography of John Herschel

GÜNTHER BUTTMANN

TRANSLATED BY B. E. J. PAGEL, ROYAL GREENWICH OBSERVATORY

EDITED AND WITH AN INTRODUCTION BY DAVID S. EVANS,

DEPARTMENT OF ASTRONOMY, UNIVERSITY OF TEXAS

Illustrated with Photographs and Drawings

LUTTERWORTH PRESS · *Guildford & London*

FIRST PUBLISHED IN GREAT BRITAIN 1974

Frontispiece: Sir John Herschel, (photograph by Julia M. Cameron, 1867)

ISBN 0 7188 2087 8

PRINTED PHOTOLITHO IN GREAT BRITAIN BY
EBENEZER BAYLIS AND SON LIMITED
THE TRINITY PRESS, WORCESTER, AND LONDON

Contents

Illustrations

Preface

To attempt to describe the extraordinarily rich and varied career of John Frederick William Herschel is to face several difficulties. Herschel's work extended over many scientific fields, each now a separate specialty. This diversity compels the treatment of many topics only by mention or reference. Others, more familiar to the author, have been treated with greater completeness. Some such unevenness is inevitable in any scientific biography but is probably unusually pronounced in this. The work is less than a complete portrait, but if the sketch succeeds in bringing to life Herschel's scientific achievements and vivid personality it will have achieved much. If it stimulates study of Herschel and his times it will have achieved more, for, nearly a century after his death, no complete biography exists. This is remarkable when one considers the commanding position Herschel occupied among Victorian scientists. Such a gap in the literature can hardly be filled by a single publication.

In the preparation of this book, I have used the following unpublished primary sources: John Herschel's diaries, 1834–1871, copies in the possession of the Royal Society, London; John Herschel's letters, 1812–1871, copies in the possession of the Royal Society, London, and extracts compiled by Mira F. Hardcastle, a granddaughter of John Herschel, and in the possession of Mrs. E. D. Shorland, a great-granddaughter; fragment of a biography of John Herschel (1792–1830) by Mira F. Hardcastle, in the possession of Mrs. E. D. Shorland; Chemical Experiments, vol. 3, notebook by John Herschel in the possession of the Science Museum, London; Record Books of the Royal Mint, containing correspondence and memoranda of John Herschel, 1851–1854, in the Archives of the Royal Mint, London. Important published sources on the Herschel family are *Memoir and Correspondence of Caroline Herschel (1750–1848)*, edited by Mrs. John Herschel, and *A Short Biography of Sir John F. W. Herschel, Bt.*, by Constance A. Lubbock, a fragment of a memoir by Herschel's youngest daughter. These books are included with other biographical works on John Herschel in the Bibliography, which also contains a selected list of Herschel's own writings.

Grateful acknowledgment for permission to use illustrations as indicated is made to the following persons and institutions: The Deutsches Museum, Munich–frontispiece; The Kodak Museum, Harrow, England–pages 84 (bottom), 87 (bottom); Patrick Moore, F.R.A.S., East Grinstead, England–page 85; The National Portrait Gallery, London–pages 76, 78; The Royal Astronomical Society, London–page 79; The Science Museum, London–page 84 (top); Mrs. E. D. Shorland, Bracknell, England–pages 75, 77, 80, 81, 82, 86, 87 (top). The illustrations on pages 83 and 88 are reproduced from J. F. W. Herschel, *Results of Astronomical Observations . . .* (London: Smith, Elder, 1847).

Günther Buttmann

Introduction

The sons of famous fathers are at a disadvantage in achieving distinction for themselves. Of none is this more true than of John Herschel, only son of William Herschel. To his contemporaries the father was notable as the discoverer of Uranus, first addition to the classical list of naked-eye planets. Mature consideration esteems him more as the founder of stellar astronomy, the first to extend systematic observations beyond the bounds of the solar system into galactic and extragalactic space, using the large telescopes of which he was equally a pioneer.

Full-scale biographies of him are numerous; so also, though more abbreviated, are the notices of his sister, Caroline Lucretia Herschel. Her personal and scientific devotion to her brother during his long bachelorhood has become legendary. She fed him with her own hands as he worked at grinding his mirrors; she became an observer and discoverer of comets in her own right. After his death she returned to her native Hanover, where she lived to be almost a hundred, celebrated as much for the lively charm of her personality as for her scientific achievements.

William married only late in life when his fame was fully matured. His position in the contemporary scene must have resembled that of Einstein, for his name was known to every educated person even though the details of his achievements might be hazy. Then son John made his appearance, an only child in an astronomical household, which kept the strange hours which so often make astronomical observers comparative strangers to their families. Aunt Caroline, forty-two years older than the solitary little boy, took him to her heart, and there began a warm relationship which lasted for more than half a century. After Caroline went to Hanover there was a voluminous correspondence between the two which reads less like that of an aunt and nephew bridging a generation gap than that of a brother and sister.

That John benefited by his inheritance there is no doubt. The father had made his way from poverty and obscurity to a state of affluence and distinction. The son was born into a world where his merits might immediately be recognized by the galaxy of savants and aristocratic patrons with whom he was brought into contact, at first through the family connections. His was a name to conjure with. On one occasion, arriving in France, he showed his documents to a *douanier* who exclaimed: "Herschel! That is not a name: it is a star!"

The advantages had to be paid for: comparisons with his father must have been inevitable. However, he was gifted with a powerful and versatile intellect which enabled him to excel as a student of mathematics at Cambridge. Ranked Senior Wrangler, that is, best of his year, in the verbal debate which then constituted the examination, he was dismissed, as he recorded, "with a flaming compliment." The study of astronomy was a family heritage which he worthily carried on, in this field inviting direct comparison with his father. In other fields, such as mathematics and chemistry, his achievements were comparable and more diverse. The father was a self-educated genius unequaled in his specialty: the son, through his formal education, knew how to spread his talents over many fields.

The implied verdict of history has been to rank the father before the son, for Günther Buttmann's is the first full-length biography of the latter. The record of John's scientific work contained in this book shows that there has been an injustice to repair and a merit inadequately recognized. However, it would be better to recognize that the merits of these two men are of different kinds, rather than to set their reputations in competition. Certainly, the latter would have been abhorrent to John, who always spoke of his father with a respect and admiration bordering on reverence.

Günther Buttmann's biography goes beyond the simple record of the amazing diversity of John's scientific contributions. To do no more than that would have been to miss the essence of the man. In spite of the hazards of his upbringing, John attained a full development of his personality, which comes to us through the historical records and documents with as strong and attractive a voice as that of dear old Caroline. The happiness of his mature years was largely due to his supremely fortunate marriage.

Engineered by a friend, occurring at an age then regarded as within sight of middle age, with a beautiful bride still in her girlhood, the union was idyllic. Physical passion and a compatibility of intellect there certainly were, and beyond these were the wife's qualities of care-free gaiety and readiness to share the hardships and dangers of the African visit. To the end of Herschel's days, his diaries continue to show his devotion to his wife and their twelve children.

There may be yet more to be learned of John's life before his marriage at the age of thirty-seven, for unpublished documents exist that hint at some previous entanglement which was disapproved of by the family and came to nothing. It would indeed be surprising if one of John's temperament and evident attraction had reached that age entirely heart-whole.

Günther Buttmann describes his book as a sketch for a biography. This modesty is justified only in the sense that there may be yet more to be written about John. What he does achieve is to draw attention to one of the liveliest minds and most attractive personalities of the nineteenth century, until now undervalued by historians of science.

David S. Evans

THE SHADOW

OF THE TELESCOPE:

A Biography of John Herschel

1

The Young Prodigy

✦

✦

✦

John Frederick William Herschel, only child of the great astronomer Sir William Herschel, was born at Slough, England, on March 7, 1792, into a home of extraordinary distinction. The household was described by the astronomer Charles Pritchard as "singularly calculated to nurture into greatness any child born, as John Herschel was, with natural gifts capable of wide development. . . . At the head of the house there was the aged, observant, reticent philosopher; and rarely far away, was his devoted sister, Caroline Herschel, whose labours and whose fame are still cognisable as a beneficent satellite to the brighter light of her illustrious brother. It was in the companionship of these remarkable persons, and under the shadow of his father's wonderful telescope, that John Herschel passed his boyish years. He saw them, in silent but ceaseless industry, busied about things which had no apparent concern with the world outside the walls of that well-known house." [1]

At the time of his son's birth Frederick William Herschel (1738–1822) was at the height of his astronomical career, which had taken him from the most humble beginnings to the pinnacle of fame. The great advantages enjoyed by his son—an excellent education which provided the opportunity for a brilliant career—combined with a carefully nurtured and wholly untroubled youth—had been lacking in the case of the father. William Herschel was the son of a simple army musician in Hanover. He adopted the same profession as a boy of fourteen. In 1758 he came to England, a penniless refugee from the confusion and misery brought upon his native city by the Seven Years' War, and for some years eked out a bare existence, first as a musical copyist and then as instructor of music to a small military band in the north of England. Finally, in 1766, he was appointed to the post of organist at Bath, at that time a famous and fashionable spa; later he also became director of the Bath Orchestra.

There in his spare time he began to read works on mathematics and astronomy and to make attempts at grinding telescope mirrors. Soon it became his dearest longing to possess a telescope of his own. "I resolved," he once wrote, "to take nothing upon trust, but to see with my own eyes all that other men had seen before."

His first homemade instrument was a reflecting (or mirror) telescope of 7-foot focal length. Its principal concave mirror was about 6 inches in diameter, and it was constructed on the Newtonian principle (deflection of the converging rays by means of a small flat mirror into an eye-piece placed at the side of the tube). William Herschel made the eye-pieces himself; the strongest one had a magnifying power of 6450. The telescope was mounted on a wooden framework with rollers. The orientation of the tube was achieved by a system of ropes and screws.

With this first telescope he undertook a survey of the whole northern sky, observing everything that came within its field of view. His interest in astronomy now became the great passion of his life. Every minute that he could steal from his job was used in grinding telescope mirrors and observing the stars.

He found faithful and enthusiastic helpers in his sister Caroline and his brother Alexander, who had joined him at Bath after the death of their father. Caroline became his astronomical assistant, while Alexander helped in the construction of telescopes. Carried along by their brother's enthusiasm, they formed an ideal partnership—a rare example of three people serving a common purpose with equal devotion. Their interest almost reached the point of obsession, so absorbed were they in their esoteric activities.

The great turning point in William Herschel's life resulted from an event that did not seem to him at the time to have any great significance. On March 13, 1781, he observed in the constellation of Gemini an object which he could not find on any celestial chart. Assuming that he had found a comet, he reported the discovery to the Royal Society of London, to which he had already communicated several of the results of his observations, in a short paper entitled "Account of a Comet," which was published in the *Philosophical Transactions* of the Society.[2]

Several observatories undertook the search for Herschel's object. As was soon discovered, it showed no resemblance to a comet, and furthermore was not describing a cometary orbit. The famous French astronomer Pierre Simon de Laplace (1749–1827) finally carried out an exact computation of the orbit on the basis of the observational data and found that Herschel had discovered, not a comet, but a new planet, the seventh in the solar system, circling around the central sun at twice the distance of Saturn.

This brought about a sensation in the scientific world such as can hardly be imagined today. From the earliest beginnings of astronomy in ancient times, only five planets —Mercury, Venus, Mars, Jupiter, and Saturn—had been known, although since the time of Copernicus earth had been counted as the sixth. The assumption that the solar system was bounded by the ringed planet Saturn was so deeply ingrained in astronomical thinking that it had never occurred to anyone to look for an additional planet. Herschel himself stumbled upon it only by accident. The new planet, which was named Uranus, had actually been observed some twenty times since 1690, but it had always been recorded on celestial charts as a fixed star.[3]

With his discovery, the hitherto unknown amateur astronomer of Bath had advanced the knowledge of the solar system into undreamed-of depths of space. Soon Herschel's name was on everyone's lips. The first reactions were enthusiastic eulogies on the one hand; skepticism, distrust, and envy on the other. But Herschel had made his mark. King George III received him at court and encouraged his activities by providing him and his sister with fixed annual salaries of 200 pounds and 50 pounds, respectively. This enabled Herschel to give up his position as orchestra director and organist and to devote all his time to astronomical researches.

In 1786 the Herschel family moved to Slough, a town in the neighborhood of Windsor, where a house with a large garden was acquired. Work was immediately begun on the installation of an observatory on the property. Supported by a considerable grant from the royal purse, Herschel undertook the construction of a huge telescope with a focal length of 40 feet and a main mirror 48 inches in diameter.

Until it was dismantled in 1839, this telescope ranked as the largest reflector in the world. The eye-piece was placed at the front end of the tube and was accessible from a movable platform. The telescope was constructed on the now obsolete Herschelian principle, in which the primary mirror was slightly inclined to the optical axis and so reflected the light rays directly into the eye-piece; this eliminated the secondary plane mirror of the Newtonian form, which obstructs some of the incident beam, and thereby somewhat increased the light-gathering power of the instrument.

After three years of constructional work, Herschel was

able to point the great tube at the sky, and the first notable success that he achieved with it was the discovery of the sixth and seventh satellites of Saturn. The enormous mounting, with its tall ladders and beams that over-shadowed all the rooftops of the little town, became not only a landmark in Slough but the symbol of a new epoch in astronomy.

William Herschel's activities now steadily increased. Day after day, with Alexander's help, he produced mirrors and telescopes which found purchasers all over Europe. At night he sat at his telescope with Caroline and observed double stars, nebulae, and star clusters, which the great light-gathering power of his telescopes enabled him to discover by the hundred. In well over 3000 areas of sky, he carried out star counts—"gauges," he called them—for the purpose of using the distribution and brightness of the stars to determine the form and spatial extent of the Milky Way system. He discovered the systematic or "peculiar" motion of the solar system through space—a brilliant piece of work which would in itself have been sufficient to ensure his permanent scientific fame. His great lists of star clusters and nebulae, along with his cata-logue of double stars, bear witness to the indefatigable industry of this extraordinary man.

Such, then, was the scientific career of the "observant, reticent philosopher" described by Pritchard. William Herschel, then in his late forties, had never married, but not long after the move to Slough he changed his state. He married Mary Pitt, née Baldwin, the young widow of a well-to-do London merchant, on May 8, 1788. The mar-riage appears to have been a happy and harmonious one. Although not endowed with outstanding intellectual gifts, Mary Herschel had a winning, radiant personality, which brought warmth and liveliness into the austere astronomi-cal household. Their son was born after four years of marriage.

The environment in which John Herschel grew up,

despite all its advantages and unique educational opportunities, had one unavoidable disadvantage; the age of his parents. (William was 54 when John was born.) John always regretted that his half-brother, Paul Pitt, died before he was old enough to know him. His cousin, Sophia Baldwin, who was part of the family, was eight years older but they were good companions. He was thus surrounded by adults whose activities he followed with shy admiration. Often when his father and aunt were resting from their night-time astronomical activities, silence was imposed on the household during the day, and noisy games were not allowed. While there was no lack of company in the Herschel house and at times guests arrived almost every day, they were mostly astronomer friends of William Herschel's. The conversations to which little John was allowed to listen were usually beyond a child's comprehension.

Serious and adult as his home was, however, John's youth seems to have been cheerful enough. Dreamy, and mature beyond his years, he may have been but he was also a normal boy. His Aunt Caroline recounted that he hung about among his father's craftsmen and builders, received instruction from them in the use of tools, and then one day set about undermining the family home with a hammer and chisel. He also seems to have terrified his parents on several occasions by daredevil climbs on the ladders and scaffoldings of the telescopes.

Caroline developed a great affection for her nephew, and the hours he spent with her in the small house to which she had moved after her brother's marriage were especially happy ones for the boy. "Many a half or whole holiday he was allowed to spend with me," Caroline wrote later, "was dedicated to making experiments in chemistry, where generally all boxes, tops of tea-canisters, pepper-boxes, teacups, etc., served for the necessary vessels and the sand-tub furnished the matter to be analysed. I only had to take care to exclude water, which would have produced havoc on my carpet." [4] It is characteristic of John Herschel that he should choose a branch of science for his childish games. He continued to pursue chemistry enthusiastically in later years, even though it did not become his principal life work.

On May 1, 1800, just before he reached the age of eight, John was sent as a boarding pupil to Eton, which was only a mile from Slough on the highway to Windsor. The Herschels could well afford to have their son educated at the most expensive, famous, and exclusive school in England, but John's period in residence did not last long. One day his mother, no doubt overanxious about his somewhat delicate health, saw her son inveigled into a boxing match by an older and stronger boy and knocked to the ground. Fearing the effect on the boy of such rough episodes, which were by no means unusual at Eton, Mary Herschel summarily withdrew him from the school. He was then sent to a private school run by his father's friend Dr. Gretton at Hitcham, a village in the neighborhood of Slough, and a private tutor was also engaged to instruct him at home with the object of preparing him to enter a university. The tutor was a Scottish mathematician named Rogers, who seems to have been a highly capable man. He not only taught his pupil the elements of the natural sciences but also introduced him to modern languages, literature, and music, thus providing a most valuable complement to the purely classical course provided by Dr. Gretton. In mathematics, however, his efforts were at first completely unsuccessful. This failure is especially surprising in view of John Herschel's subsequent career: he was made a Fellow of the Royal Society at the early age of twenty-one as a result of a brilliant mathematical investigation; a few years later he made a superb contribution to the introduction of continental methods of mathematical analysis into England; and in 1821 was awarded the Copley Medal, again for a series of mathematical papers. After a short time, however, Rogers' instruction seems to have yielded excellent results. There is an interesting parallel here with Herschel's great fellow countryman Sir Isaac Newton (1642–1727), who is also reported to have been a poor mathematician at school.

During the summers of the pre-university years, John was allowed to accompany his parents on some major journeys. His father wished to show the boy, who had grown up in undue isolation at Slough, something of the world into which he would soon be entering on his own account. They traveled in their carriage round England, Wales, and Scotland, visiting in turn the towns in which William Herschel had led a hard and poor but also a happy life as a young musician. There was even a trip to Paris, where John was shown the sights by his godfather Count Komarzewskī, while his father, accompanied by Laplace, had an audience with Napoleon.

In October 1809, at the age of seventeen, John Herschel took up residence at the University of Cambridge, where he matriculated as a student of mathematics and physics in St. John's College. The beginning of his university career marked a decisive step in his intellectual development. A wealth of new impressions overwhelmed him. The quiet, somewhat dreamy boy from the astronomer's household in the little country town of Slough was probably bewildered at first by the cosmopolitan university town. But he soon found a stimulating circle of friends in the relatively small and intimate study group at St. John's. He developed particularly close friendships with George Peacock (1791–1858), a mathematician who later became a theologian and Dean of Ely, and with another mathematician, Charles Babbage (1792–1871), who was to devote most of his life to the invention of a calculating machine.

Every Sunday morning after chapel the three friends were accustomed to spend a few hours together in lively scientific and philosophical discussion. With youthful and romantic ardor, they sealed their friendship with a resolution "to do their best to leave the world wiser than they found it."

The first fruit of their exacting program was the Analytical Society of Cambridge, which they founded in 1812. The object of this society was to make known in England the modern methods of infinitesimal calculus that had been developed chiefly in France and Germany and to replace the rather cumbersome notation of Newton's "calculus of fluxions" by the more elegant usages practiced on the Continent. This ambitious project of the three undergraduates was to achieve remarkable success within five years.

The task which the Analytical Society had set itself was revolutionary, not only from a purely mathematical point of view, but from that of the history of thought in general. There was probably no university in England where the supreme authority of Newton was more completely and devotedly maintained than at Cambridge, where Newton himself had taught. His *Principia*[5] provided the basic apparatus of every student of mathematics and natural sciences. John Herschel himself, in order to enter completely into the spirit of the work, is reported to have cast aside the English translation normally used by students and read the *Principia* in the original Latin, no mean linguistic achievement in view of the forbidding character of seventeenth-century scientific Latin. Newton's authority was held to be unassailable, even in fields having a more speculative and philosophical content, such as the theory of light.*

---

* Newton supposed that light consisted of minute material particles which are expelled by the source and propagated in straight lines in all directions, but the emission theory was much more sophisticated than it is often represented to be and included features of a very modern flavor, such as the probability notions of "Fits of Easy Reflection and Transmission."

This veneration of Newton, reinforced by the strong traditionalism in English universities and by immense national pride, had prevented the adoption of the analytical methods of Newton's contemporary, the German mathematician and philosopher Gottfried Wilhelm von Leibnitz (1646–1716) or those of Laplace. Attempts made at the beginning of the nineteenth century by two British mathematicians, Robert Woodhouse (1773–1827) and James Ivory (1765–1842) to introduce these methods into English science achieved no success, although Woodhouse's textbook on trigonometry[6] was well received and had a marked influence on Herschel. The revolution seems to have required the infectious youthful dynamism of Herschel and his friends.

Herschel and Peacock translated *Traité du calcul différentiel et du calcul intégral* by the French mathematician Sylvestre François Lacroix (1765–1843).[7] The translation, which appeared in 1816, was enthusiastically received by the younger generation of mathematicians and college lecturers. Soon it became generally adopted as a university textbook. Herschel and Babbage supplemented it with two volumes containing examples, published in 1820; one of them, written by Herschel, contains a wealth of material on the calculus of finite differences.[8] Questions based on the new continental methods introduced by the Analytical Society were used in the examinations held at Cambridge in 1817 and 1819, and these methods were accepted all over England in an amazingly short time, displacing Newton's fluxions.

Their struggles for mathematical reform in no way reflected an intention to belittle the supreme scientific greatness of Newton, which Herschel and his contemporaries fully recognized. Herschel's burial next to Newton's grave in Westminster Abbey, half a century after these events, was more than merely a pious gesture.

Herschel's student years at St. John's were marked by

an uninterrupted series of successes. Caroline Herschel, overflowing with almost childish pride in her nephew, wrote in her diary that "from the time he entered the University till his leaving he had gained all the first prizes without exception." In *Nicholson's Journal*, in February 1812, Herschel published anonymously a collection of analytical formulas,[9] which was the first printed result of his introduction of new methods of mathematical analysis; this was followed shortly afterward by a compilation of trigonometric formulas in the same periodical.[10] In October of the same year John Herschel submitted to the Royal Society, through his father, a mathematical paper "On a remarkable application of Cotes's Theorem." [11] This paper, inspired by reading Woodhouse's textbook, aroused much admiration by the elegance of its argument. On May 27, 1813, its author was elected a Fellow of the Royal Society—a most unusual honor for a young student. This was a precursor of three later papers on analytical topics, published in *Philosophical Transactions* in 1814, 1816, and 1818,[12] which won for Herschel in 1821 the Copley Medal, the highest scientific award bestowed by the Royal Society.

The year 1813 was a successful one for Herschel in another respect. In January he took the tripos (the university examination in mathematics) for his Bachelor of Arts degree and was classed Senior Wrangler (best candidate) and Smith's Prizeman. Peacock only managed second place, while Babbage voluntarily withdrew, having seen that he could not successfully compete with Herschel. This incident, however, caused no breach between Herschel and Babbage. Apart from the personal affection in which each held the other, their common preoccupation with the new analytical ideas formed a strong bond between them. They exchanged letters the length of respectable scientific papers on mathematical problems, the solutions to which Herschel developed with such facility and elegance as to make them seem completely obvious. Plans were laid for the future of the Analytical Society, statutes[13] drafted, and members elected. The first volume of *Memoirs of the Analytical Society* came out, also in 1813, and included an extensive contribution by Herschel on differential equations and their applications.[14]

With this series of publications, the society hoped to win supporters outside the confines of the university and thus to introduce the new methods of analysis into the studies and lecture rooms of British mathematicians and scientists. In a letter to Babbage in 1813 Herschel writes, "For I repeat it again and again: we must not be a *Cambridge* Analytical Society." [15]

In addition to his mathematical studies, Herschel now turned to chemistry. Whether he was stimulated by Babbage, who had taken up chemistry after "migrating" to Trinity College, is not clear from the two friends' correspondence. Herschel installed his own small laboratory in his father's house at Slough, where he spent his vacations. Full of enthusiasm, he plunged into the new science, which at that time was undergoing profound changes as a result of fundamental research by the English chemist Sir Humphry Davy (1778–1829), the French chemist Joseph Louis Gay-Lussac (1778–1850), and others. Aware of the limitations imposed on his creative drive by the amount of time available, if by nothing else, Herschel once wrote to Babbage: "God knows how ardently I wish I had ten lives, or that capacity, that enviable capacity, of husbanding every atom of time, which some possess, and which enables them to do ten times as much in one life. . . ." [16]

John Herschel was now at the crossroads of his intellectual development. He hesitated whether to devote himself to a purely scientific career or to take up a more lucrative occupation and restrict his scientific pursuits to his leisure. He decided in favor of a gainful occupation, feeling that, despite his preference for mathematics, his

scientific interests were so manifold, urgent, and frag-
mented that it would have been difficult for him to com-
mit himself to any one particular field. His restless spirit
required a high degree of freedom of movement. Yet he
may have felt the need for the firm line of duty which is
the inevitable concomitant of any profession, as a healthy
counterpoise to the dangers of the very diversity of his
interests. He was perfectly well aware of these dangers, as
various remarks to his friends show. His extraordinary
intellectual adaptability, his limitless capacity for enthusiasm at a new idea, and the almost hectic restlessness of
his scientific efforts remained characteristic even in old
age. These traits were a part of his heritage from his
father, but in William Herschel they were associated with
the essential qualities of a scientific pioneer and revolutionary thinker, qualities that were less prominent in the
son.

In January 1814, Herschel wrote to Babbage[17] that he
had decided to become a lawyer and intended forthwith
to move to London and read for the bar at Lincoln's Inn
(one of the four legal societies of London through which
young lawyers receive their training and their admission
to the bar). This decision seems to have been made very
suddenly and against the wishes of his father. The elder
Herschel would have liked his son to become a clergyman,
a desire that seems very strange if one makes the reasonable assumption that his greatest hope would have been
for John to be an astronomer and continue and complete
his own researches. However, the wish seems to have
arisen, not from religious feeling or any basic preference
for a clergyman's calling, but from the somewhat superficial and indeed questionable assumption that clerical
duties would provide more leisure for the pursuit of private hobbies and scientific interests than any other profession could offer.

John, however, did not relish the prospect of this peace-

ful and narrow existence, although his father presented it
in the most glowing colors in a long letter. In spite of his
father's low opinion of the legal profession he finally
received permission to enter it. He began to read for the
bar in February 1814, and was introduced to the practical handling of cases in the chambers of a Mr. Sander.

From the very beginning Herschel does not seem to
have been happy in this new and radically different environment. His regular attendance at meetings of the
Royal Society continually drew his interest back to science,
the more so as he became increasingly aware of the emptiness of the lawyer's existence as it was presented to him
at Lincoln's Inn. His distaste appears in his letters to James
Grahame,* a friend he first met in Glasgow, although he
wrote to Babbage: "I am determined, as the profession is
of my own chusing, much against the wish of my parents,
that I will pursue it in good earnest." [18]

Certainly he never felt any genuine love for the law;
instead, his scientific interests were only strengthened
during his stay in London. A decisive influence was his
acquaintance with the chemist William Hyde Wollaston
(1766–1828), whose fascinating lectures revived all his
former interest in chemistry.

During Herschel's London period he also made the acquaintance of the astronomer Sir James South (1785–
1867), with whom he was later to carry out a systematic
revision and continuation of William Herschel's astronomical observations. South, who came from a well-to-do
family, had studied medicine and practiced as a doctor

* James Grahame, sometimes described as a "historian" and known as
a matchmaker friend of both Herschel and his future wife's family (see
Chapter 3) and as the author of verses, is a slightly mysterious figure.
He may have been one of the sons of the Scottish poet the Reverend
James Grahame, who died in Glasgow in 1811, and the James Grahame
who entered St. John's College from Glasgow in 1811 at the age of
twenty. The background in each case fits very well. Nothing is at present
known of Grahame's later life except that he remained friendly with the
Herschels.

for a few years, but at the time John Herschel met him he was devoting himself entirely to astronomy and had installed a private observatory with excellent instruments in his London house in Blackman Street.

Soon Wollaston and South completely deflected Herschel from a legal career, and his eighteen months at Lincoln's Inn were a period of intensive studies in science rather than in law. In March 1815, he applied for the Chair of Chemistry that had just become vacant at Cambridge, but his rival was elected by a majority of one vote. Resignedly he wrote to Babbage: "I made my retreat with as good a grace as might well be and have nothing for my pains but the satisfaction of knowing that if Tennant [the previous occupant of the Chair] had lived a twelvemonth or two years longer I should in all probability . . . have been his successor. I do not care much about it and on the whole I believe it is better as it is.—I have been attending Clarke's* lectures and am become half a mineralogist—and have been analysing some of his specimens for him. . . ." [19]

Herschel's legal career came to an abrupt end in the summer of 1815. The exertions demanded by his double course of studies and by his restless mind, and the unaccustomed pressures of life in the metropolis, had undermined his health, never very robust, and he required medical attention. Several weeks at the seaside resort of Brighton restored his vigor, but at the beginning of November he wrote to Babbage: "Indeed I begin to fear I must throw up the profession altogether as a lost game and I endeavour to reconcile myself to it. . . ." [20]

He did not return to Lincoln's Inn, although it was not until a year later that he had his name struck from the register. After lengthy consideration, he accepted an offer that had been made to him in May 1815 by one of his former teachers at St. John's, the mathematician the

---

* Edward Daniel Clarke (1769–1822), English mineralogist.

Reverend Thomas Waldron Hornbuckle (1775–1848). The post was that of sub-tutor and examiner in mathematics at St. John's and, though modest, did at least offer the opportunity of an academic career. Herschel had originally declined the offer because at that time he still had the firm intention of preparing for the bar.

It was not difficult for him to take leave of the legal career which he had entered upon halfheartedly and purely for considerations of prudence. Once more and with redoubled enthusiasm, he plunged into his mathematical researches and renewed old friendships, especially that with his friend of undergraduate days, William Whewell (1794–1866), the gifted mathematician, physicist, and philosopher who later became Master of Trinity College. Whewell, undoubtedly one of the most remarkable men at Cambridge University in the first half of the nineteenth century, was, like Babbage and Peacock, an enthusiastic supporter of the new methods in mathematical analysis. His *Textbook of Mechanics*, which was imbued with the spirit of modern analysis, appeared in 1819. Herschel's warm friendship with Whewell lasted until the latter's death.

The year 1816 began with a phase of intensive work for Herschel. His duties as sub-tutor were exacting and somewhat monotonous. He complained to Babbage: "You are pretty well aware what a job it must be to be set from 8 to 10 or 12 hours a day examining 60 or 70 blockheads, not one in ten of whom knows his right hand from his left, and not one in ten of whom knows anything but what is in the book. . . . In a word, I am grown fat, full and stupid. Pupillizing has done this—and I have not made one of my cubs understand what I would have them drive at." [21]

This disappointment did not impair his own enthusiasm for mathematics. The translation which he and Peacock made of Lacroix's *Traité* was carried out in this period of "pupillizing," as was the preparatory work for the collection of examples on the method of finite differences published in 1820 as a supplement to the translations of Lacroix.

On July 3, 1816, Herschel took the degree of Master of Arts and was elected a Fellow of St. John's College—that is, a member of the academic faculty. The end of his years of study and wandering appeared to have been reached, and a promising academic career seemed to lie before him. His scientific publications, his leading role in the Analytical Society, and his Fellowship in the Royal Society had already established for him a reputation that was not merely a reflection of his father's achievements but had a brilliance and character of its own. But new circumstances now intervened to turn his life in a completely different direction.

In the summer of 1816 he accompanied his father on a trip to Dawlish, a popular resort on the Devonshire coast. William Herschel, who was then in his seventy-eighth year, had for some time suffered from various ailments brought on by advancing age and had been compelled to restrict his astronomical observations more and more. Caroline also could no longer bear the strain of long nights of watching at the telescope. However, William Herschel could not make up his mind to give up his astronomical researches entirely. After forty years he had become so involved in his work that he could not tear himself away from it. Important parts of his life's work were still incomplete, while others required revision, reevaluation, and systematic collation. The task was too great for the aged astronomer to cope with alone, and the choice of a successor who would continue and complete it, in his spirit and using his methods, pressed on him with ever-increasing urgency.

Father and son must have held many earnest conversations on these matters. In the end, John Herschel offered to become his father's astronomical assistant. This decision, arising out of love and respect for his father in the last stage of his restless life, started John Herschel on what ultimately became his own life work, despite all his other, often mutually contradictory interests. The choice must have been made with great hesitation and uncertainty and could not have been motivated by either romantic enthusiasm or cold calculation, but nevertheless it was to lead to achievement and fortune.

Parting from Cambridge—unlike leaving the law—was not easy. Although at the time he had not decided to give up forever the career he had begun, he seems to have sensed that the departure was to be final. On October 10, 1816, he wrote to Babbage: "I shall go to Cambridge on Monday where I mean to stay but just time enough to pay my bills, pack up my books and bid a long—perhaps a last farewell to the University. . . . I always used to abuse Cambridge as you well know with very little mercy or measure, but, upon my soul, now I am about to leave it, my heart dies within me. I am going, under my father's directions, to take up the series of his observations where he has left them (for he has now pretty well given over regularly observing) and continuing his scrutiny of the heavens with powerful telescopes. . . ." [22]

2

Versatility,

Vocation, and Travel

✦
✦
✦

John Herschel's unselfish decision to give up his university career in order to assist his aging father did not mean that he became an astronomer overnight or that he gave up his former interests. These continued to hold an essential place in his life; in fact he never became an astronomer in any exclusive sense, although his principal scientific achievements lay in that field. Throughout his life his interests covered a wide range, and to some fields he returned again and again to make important contributions.

As his father's astronomical assistant, John was first taught not only the practical techniques of observing with telescopes but also those of grinding and polishing telescope mirrors. William Herschel had made innumerable experiments in the production of optical mirrors; 2160 attempts are recorded in four manuscript volumes which have not been published. Presumably his son owed some of his later knowledge to a study of these "recipe books" as well as to his father's tuition.

The younger Herschel also became thoroughly acquainted with the practice of the so-called sweeps of the heavens, which his father had developed to a high degree of accuracy. These "sweeps" or surveys were carried out by systematically observing all noteworthy objects—star clusters, nebulae, double stars, and so forth—in successive zones of the sky; the results were recorded in a catalogue. From personal experience the son came to appreciate the incredible physical exertions that the father's years of night watches had entailed.

For some decades, the study of double stars had been an important part of William Herschel's work. He had listed well over 800 of these in several major catalogues. In a number of cases he had proved that they must be binary stars—that is, physically related pairs orbiting around a common center of gravity in accordance with the law of mutual attraction between masses and not merely unrelated pairs seen by accident in almost the same direction in the sky. John Herschel continued these observations and began a new survey of the objects discovered and catalogued by his father. The purpose of these observations was to detect any changes that might have taken place in the positions of the components of these binaries and to use these as a basis for the determination of orbits. This work was continued later in collaboration with James South.

The double-star observations formed the starting point of John Herschel's astronomical career. For the time being they were only occasional exercises carried out on clear, moonless nights, while other scientific interests still took precedence. Wollaston's London lectures and the optical experiments with crystals made by the Scottish physicist David Brewster (1781–1868) had deepened his interest in the theory of light and turned his attention to the phenomenon of polarization and other optical effects. He embarked on an investigation of the optical properties of certain crystals and observed the behavior of monochromatic light propagated through crystalline substances, such as quartz, apophyllite, and Iceland spar. These crystals show a number of remarkable properties which arise from the interaction of the transmitted light with the highly ordered molecular lattice of the crystal.

One property, birefringence, is often shown for light rays falling in particular directions with respect to the crystal axes. This means that there is not one ray but two emerging from the crystal for every ray that enters, and an image seen through a birefringent crystal will not be single, as it would be if the crystal were replaced by a similarly shaped piece of glass, but double.

One ray is called the ordinary ray because its behavior is like that found for a crystal-shaped slab of glass. The other ray, the extraordinary ray, obeys different rules. In particular, the phenomenon of polarization occurs. In its simplest form this means that the waves in a particular beam of light are all oscillating in the same plane. Certain substances will transmit only light polarized in a particular plane. A very good demonstration of this is given by Polaroid sunglasses, which cut off the polarized component of the sunlight produced by the atmosphere which is most marked in directions in the sky perpendicular to the sun when the glasses are correctly turned in their own plane.

In an extensive paper published in 1820,[1] Herschel described an experimental arrangement to demonstrate polarization with the aid of two tourmaline plates. If the crystal axes of the two tourmalines are perpendicular to each other, the extraordinary ray transmitted by the first plate does not penetrate the second. Herschel's discovery of the relationships between the crystalline structure of a transparent substance and its optical properties was described by the distinguished Scottish physicist William Thomson, Lord Kelvin (1824–1907), as "one of the most notable points of meeting between Natural History and Natural Philosophy."[2]

In the case of apophyllite, Herschel discovered that blue light underwent positive birefringence and red light negative birefringence, while green light was simply refracted. He reported this in the *Transactions of the Cambridge Philosophical Society* for 1821.[3] Another paper, discussing the absorption of light in its passage through glasses, liquids, and other substances, was published in 1823 in the *Transactions of the Royal Society of Edinburgh*.[4] Herschel had been elected a Fellow of that society in 1820. The experiments described in this paper are of some importance in the early history of spectrum analysis. Herschel observed a great variety of spectra of flames and carefully investigated the bright emission lines.

A simple explanation of these terms may be helpful here. One of the most powerful of optical instruments is the spectroscope, in which light of many colors, all mixed, coming from a star, the sun, a lamp, or a hot body, is separated into its individual rainbow colors by a prism or a more sophisticated device called a diffraction grating. The light as dispersed in this way is called a spectrum, and the instrument which does this is called a spectroscope or spectrograph. In principle, each point in the spectrum corresponds to a particular color or wavelength of light, and the brightness of the spectrum at the corresponding point shows how much or how little of this color of light there was in the original mixture. To make the spectrum pure, the incoming light is usually admitted through a narrow slit, and lenses or mirrors are used, so that the spectrum becomes a series of sharp images of the slit set side by side in a row, each image being formed in light of just one sharply defined color. If a color is missing or weak, there will be a dark absorption line in the spectrum at the corresponding position. If a color is present in excessive abundance, the spectrum will be very bright at a given position, and there will be an emission line. The enormous usefulness of spectrum analysis arises from the fact that each chemical element has a precisely characteristic pattern of spectrum lines. In general, if the element is heated to a temperature sufficient to vaporize it, the lines from the emitted radiation will show as emission lines. If the vapor transmits light from a still hotter source of radiation behind, the spectrum will show absorption lines.

From the position of the lines in the continuous spectrum Herschel was able to derive the chemical composition of the source of light. His studies revealed the possibility of making chemical analyses through observation and interpretation of the spectrum. About forty years later, the German physicist Gustav Robert Kirchhoff (1824–1887) discovered the explanation of the dark absorption lines that appear in the spectrum of the sun and, with the German chemist Robert Bunsen (1811–1899), laid the foundations of spectrum analysis. Kirchhoff showed that dark lines occur because of the presence, between the light source and the spectroscope, of cooler gases which absorb light at just those wavelengths that they would themselves emit when incandescent. In his paper on the solar spectrum,[5] Kirchhoff showed that the absorption lines are caused by a cooler gaseous envelope surrounding the incandescent globe of the sun. The spectroscope could thus be used to examine the chemical composition of the sun, and in principle that of any other luminous heavenly body which presented a recognizable spectrum. Herschel was not in a position to recognize the real cause of the absorption lines because he accepted his father's theory that the sun was a dark body surrounded by a luminous envelope.[6]

John Herschel's interests extended also to other effects in the field of physical optics. He described the interference of light as produced by the layered structure of mother-of-pearl,[7] a phenomenon which depends on the wave nature of light and arises by interaction with a regular structure of a size comparable with the wavelength. At that time the nature of light was being actively studied by scientists. The French physicist Augustin Jean Fresnel (1788–1827), stimulated by the ideas of the English physicist Thomas Young (1773–1829) and the French physicist Dominique François Arago (1786–1853), was carrying out the mirror experiments that provided a firm basis for the wave theory.

Herschel also invented a simple arrangement to demonstrate the interference of sound waves. This apparatus, named after the German physicist Johann Gottlieb Christian Nörrenberg (1787–1862), consists of a glass tube divided into two arms. The two arms differ in length by half a wavelength, and interference occurs—that is, the two waves annihilate each other.[8]

An important contribution in the field of practical optics was made by Herschel in a paper on the spherical and chromatic aberrations of compound lenses,[9] which had been used since the British optician John Dollond (1706–1761) introduced the achromatic objective.

The general idea conveyed by these technical terms is not difficult to grasp. Everyone knows that a lens, whether it is a simple magnifier or a many-component lens such as is used in a camera, has the function of forming an image. The quality of this image, especially in the case of a color transparency which has to look good even when projected on a screen at enormous magnification, has to be extremely high. The difference in complexity and cost between the simple magnifier and the camera lens is a measure of the extent to which the aberrations or defects of image formation are corrected by subtle choices of shapes, positions, and materials of the lens components. Spherical aberration is a common defect of simple lenses. Chromatic aberration is a defect produced by the fact that a very simple lens has a different magnification for light of different colors. Achromatic lenses are ones in which the color effects have largely been corrected. Aplanats are lenses in which spherical and other aberrations have largely been corrected.

Herschel's calculation of achromatic aplanats, especially for microscope objectives, was carried out on the basis of a general theory of aberration using numerous examples. This paper was an attempt to develop a mathematical theory of aplanatic objectives that was addressed to practical opticians rather than to learned theorists and was to serve them as a working basis in the production of lens systems. A summary of mathematical rules for the calculation of achromats appeared in the *Edinburgh Philosophical Journal* in 1822.[10]

Even a cursory glance at Herschel's scientific publications during this period shows that their subject matter lay mainly in the realm of optics. "Light was my first love," he once said; he returned many times to optical experiments and the discussion of optical problems.

Herschel's predilection for physics went hand in hand with the almost equally strong inclination to chemistry which had shown itself even in his childhood. In 1819 he made a seemingly insignificant chemical discovery, the full importance of which was not recognized by himself or anyone else for twenty years. In two papers on the compounds of hyposulphurous [thiosulphuric] acid, he discussed, among other things, the fact that sodium thiosulphate has the property of dissolving silver salts rapidly and completely.[11] These papers received little attention at the time, but if they had reached the right audience, the invention of photography might have been achieved twenty years earlier than it was. As is described in Chapter 6, Herschel's sodium thiosulphate was to prove a decisive factor in that invention, for it provided a method of protecting the latent image produced by light rays on a layer of silver salts from destruction by the further action of light.

John Herschel also found time for mathematical research. Through his interest in optical experiments he had become acquainted with David Brewster, who was vice-president of the Royal Society of Edinburgh and had made a distinguished name for himself in the field of experimental optics. Since 1808 Brewster had been editing the *Edinburgh Encyclopaedia,* and in the summer of 1817 he approached Herschel with a request to write the contributions on the history of mathematics and on the isoperimetric problem, which had been posed by the Swiss mathematician and physicist Daniel Bernoulli (1700–1782) and examined by the Swiss mathematician Leonhard Euler (1707–1783).

Herschel readily agreed.[12] While at Cambridge he had once expressed to Babbage a desire to contribute to one of the various encyclopedias that then had great prestige in England, in accordance with the scientific-*cum*-literary spirit of the age. Herschel seems to have expected the task of preparing his contributions to be considerably more congenial than it actually proved to be; he complained to Babbage in 1818 about the difficulties in gaining access to sources and the tediousness of such historical investigations.[13] At this period his scientific interests still lay entirely in the field of experimental research, though later he was to devote much effort to descriptive encyclopedic expositions, of which he ranks as the undisputed master in mid-nineteenth-century England.

In addition to these numerous scientific investigations, John Herschel faithfully carried out his commitment to his father's astronomical researches. Many a night was spent by father and son at the telescopes, observing, writing, computing, and measuring. The first two astronomical contributions by John Herschel were published in 1822. One describes a new method of calculating occultations of stars by the moon;[14] the other is a longer paper giving tables to facilitate the determination of the places of fundamental stars.[15]

An indication that systematic astronomical work had begun in earnest is provided by a letter to Babbage in 1819: "I am sorry I cannot accompany you to Deptford on Wednesday, being engaged to pass the morning with South, in digesting (with his carpenter Mr. Flint, who possesses in perfection the organ of constructiveness) the plan of the equatorial observatory and gallows to be erected in Blackman Street . . ." [16]

Herschel's relationship with South, begun during the law student days in London, had developed into a fruitful collaboration. Herschel possessed a large amount of observational material collected by his father and also had some personal experience as an observer of double stars, since he had been carrying out these observations, with some interruptions, since 1816. Moreover, the very name of Herschel lent a certain tone to the enterprise that the two friends were planning. South's considerable personal fortune enabled him to acquire the best optical instruments that were being produced in England at the time. His two refractors (or lens telescopes), equatorials of 5-foot and 7-foot focal lengths respectively, were excellent instruments by virtue of their mechanical mountings, measuring apparatus, and graduated circles, as well as in their optical quality; they were distinctly superior to William Herschel's rather cumbersome reflectors (or mirror telescopes), which were hardly usable for accurate measurements. Moreover, the instruments at Slough had deteriorated. The great 40-foot reflector, which had been admired as a technical masterpiece after its completion in 1789, had fallen into disuse several years earlier, because the necessary repolishing of its gigantic 1-ton mirror was an extremely difficult and time-consuming operation. The 20-foot reflector also could no longer be readily used. The wooden structure of its mounting had become rotted and unsound over the years, and when John Herschel tried to reset it on the meridian (that is, pointing due south), the whole framework broke into pieces. The construction of a new telescope was undertaken at once, but even this instrument was of little use for Herschel's and South's purposes, because it did not have graduated circles of adequate precision. Like all William Herschel's telescopes, it was intended primarily as an instrument for observation. As such, it was still to serve admirably, especially in John Herschel's later survey of the southern sky at the Cape of Good Hope.

The task that John Herschel and South had set themselves was to make a new set of measurements of the double and multiple stellar systems discovered by William Herschel and other observers. The purpose was to detect any changes that might have taken place in the relative positions of the components and to use these changes as a basis for the determination of double-star orbits. The work was thus a logical extension of that begun by William Herschel with his great double-star catalogues. Almost 40 years had passed since his first observations, an interval sufficient to confirm relative motion of many pairs, even if earlier observations had been inaccurate. William Herschel had already detected movement in many of his binaries after 25 years—that is, after the lapse of about half the interval of time that John Herschel now had at his disposal.

In William Herschel's time, double stars provided one of the most popular fields in practical astronomy, largely because the subject had the charm of novelty. Any motions taking place outside the solar system aroused lively interest because they bore on the question of the applicability of Newton's law of gravitation to the realm of the fixed stars and the reduction of the dynamics of the entire universe to the known principles of celestial mechanics. The first step toward an answer had been taken by William Herschel in 1803, in a paper published in *Philosophical Transactions*.[17] He reported on six binary stars for which he had established relative motions of the components and gave the orbital periods of the companions around their primaries. A similar investigation dating from the following year includes fifty double stars for which he had detected relative motion. In this way he became the discoverer of the physical nature of binary stars—that is, he had proved that there were star pairs which form systems comparable to that of the earth and moon or to the solar system. It was his firm conviction that the force that held the two components of a binary in their orbits must be gravitation, though it was not until 1827 that the French astronomer Felix Savary (1797–1841) at Paris succeeded in proving that the orbit of the double star ξ Ursae Majoris could be represented in accordance with the law of gravitation by an elliptical orbit with a revolution period of 58¼ years.

Such orbital calculations were subject to a high degree of uncertainty because of the small amount of observational data available; for this reason it was necessary to accumulate as many measurements of double stars as possible so as to compute their orbits.

The two refractors which Herschel and South had available for their work were equatorially mounted, according to the principle of the so-called English mounting. The hour axis, which is pointed at the celestial pole and rotates to enable the telescope to follow the diurnal motion of the heavenly bodies, was constructed in the shape of a fork and its ends rested on two piers standing exactly in the plane of the meridian. The telescope was mounted on a second axis between the two prongs of the fork so that it could also be freely moved in declination (that is north and south). Two graduated circles, made by the famous telescope builder Edward Troughton (1753?–1835), from which the declination and hour-angle (east-west direction) could be read off with the aid of microscopes, enabled the telescope to be set precisely on any point of the heavens defined by the two coordinates. The objective of the smaller of the two telescopes, a Dollond achromat, had a diameter of 3.75 inches and a focal length of 5 feet, that of the larger was 5 inches across with a 7-foot focal length. In John Herschel's opinion it was one of the best telescope objectives then in use. The instruments were equipped with eye-piece micrometers* especially suitable for double-star measurements, since they could be used to determine both angular separation and position angle, the latter with a precision of 1 minute of arc.

Herschel and South began their measurements in March 1821, and continued them, with interruptions, until the end of 1823. In this period of less than three years they produced a catalogue of 380 double stars. The objects were arranged in order of right ascension; † this represents a considerable advance over William Herschel's lists, which only give the numbers of the adjacent stars in the *Atlas Coelestis* of John Flamsteed (1646–1719; first Astronomer Royal). Each double star is briefly designated by the color and brightness of the components; then follow the measures carried out by both observers, from which average values of position angle and separation ‡ are derived. Finally they give for almost all objects data

---

* Micrometers are instruments used for measuring angles in the field of view of a telescope. Cross-wires, mounted in a frame, are placed in the focal plane of the objective. A second, movable wire is shifted parallel to one of the two mutually perpendicular fixed wires by means of a fine screw. The whole system of cross-wires can be rotated about the optical axis of the telescope so that both the separation and the position angle of the two components of a binary star can be determined.

† Right ascension is the angle between the meridian passing through a heavenly body and another meridian passing through the equinox point (where the celestial equator intersects the ecliptic). Right ascension is measured from west to east and is expressed either in time units (1 hour = 15°) or in angle (from 0° to 360°). It corresponds to° longitude on the earth's surface.

‡ Position angle is the angle between an imaginary line drawn from the brighter component of a binary to the fainter component and another line starting from the brighter component and pointing due north. Separation is the angular distance between the two components of a binary, normally expressed in seconds of arc (″).

on the measures obtained by other observers, especially by the German astronomer Friedrich Georg Wilhelm von Struve (1793–1864), together with a brief judgment as to whether any changes have taken place in each case. Herschel and South were ambitious to do more than produce a double-star catalogue as a simple list. They sought to bring together all available results of observation in a kind of natural history and synopsis of double stars from which it would be possible to determine their orbits. Herschel's last great work, unfortunately never finished, was a complete catalogue of all known double stars, entitled "A general history of double stars," and containing all the verifiable measures obtained for each object since its discovery.

Herschel's and South's catalogue appeared in the *Philosophical Transactions* in 1824.[18] It won them the Gold Medal of the Astronomical Society, and in 1825 the Paris Academy of Sciences awarded them the Lalande Prize for astronomy, an honor enhanced by the presence of such great names as those of Struve, the German astronomer Friedrich Wilhelm Bessel (1784–1846), and the English astronomer John Pond (1767?–1836) on the list of candidates.

Unfortunately, the astronomical collaboration between Herschel and South came to an end with the completion of the double-star catalogue. In 1825, South went to France and continued his observations of double stars with his 7-foot refractor at Passy, near Paris, on his own, after having acquainted himself with the continental observatories and visited Struve at Dorpat. South published the results of his observations in the *Philosophical Transactions* in the form of a further double-star catalogue, containing altogether 458 objects.[19]

Another enterprise of John Herschel's early astronomical period in which South played a part was the foundation of the Astronomical Society in January 1820. The Royal

Society, a learned society chiefly distinguished for its activities in mathematics and the sciences, had existed in London since 1660. It held regular meetings at which its Fellows gave lectures on their scientific researches and discussed them. The researches were published in the Royal Society's journals, the *Philosophical Transactions* and the *Proceedings*, and thus made accessible to the scientific world. The increasing specialization and expansion of the sciences in the early nineteenth century created a need for the foundation of more specialized societies for the maintenance and encouragement of individual scientific disciplines. In 1816 a plan to start an astronomical society had been made, but the project was not realized until four years later.

On January 12, 1820, some of its supporters, including Herschel, Babbage, South, H. T. Colebrooke, the Reverend William Pearson (1767–1847), and Francis Baily (1774–1834), a London banker and a childhood friend of Herschel's, held a meeting at which the society was founded and named the Astronomical Society. In 1831 by royal charter it became the Royal Astronomical Society. John Herschel composed a foundation address which was sent to all interested persons in Great Britain and abroad. Edward Adolphus Seymour, eleventh Duke of Somerset (1775–1855), was elected president, Pearson assumed the office of treasurer, Babbage and Baily served as secretaries, and John Herschel took up the duties of foreign secretary, which consisted of creating and maintaining relations with the representatives of astronomy abroad. (Both Herschel and Colebrooke later served as president.)

The society flourished rapidly. From 1822 onward, a volume of *Memoirs* containing major scientific papers by its Fellows came out every year, while from 1831 onward shorter articles were published in *Monthly Notices*. Particularly valuable investigations were rewarded with prizes in the form of medals and "testimonials." Grants and subscriptions enabled an impressive number of excellent instruments to be accumulated. For example, a considerable grant by Pearson provided for the purchase of a 12-foot refractor with an aperture of 6¾ inches. This lens was the largest in any telescope in England at the time.

However, the Astronomical Society had some difficulties to contend with early in its existence. Its chief antagonist was the Royal Society, whose president of long standing, the naturalist Sir Joseph Banks (1743–1820), regarded the foundation of a specialized astronomical society as a threat to the unity and progress of the Royal Society. Banks prevailed upon the Duke of Somerset to give up his presidency of the Astronomical Society after a few weeks and to announce his resignation from the Fellowship. The hostility of so prominent a person as Banks made it difficult to fill the vacancy.

At last, after lengthy consultation, the aged William Herschel, vice-president of the Society, was approached with a request to take up the presidency. At first he refused, feeling that in his eighty-second year he was no longer capable of undertaking the duties of such an office. A more weighty reason, however, may have been the desire not to strain his friendly relations with Banks, which had existed for decades. The situation was changed, however, by the death of Banks in June 1820, and William Herschel was approached once more with an assurance that the presidency would not involve him in any duties, and that only nominal service was expected of him. To this he agreed, and he was unanimously elected president of the Society in February 1821.

William Herschel's term of office was to be only a short one. On August 25, 1822, after a brief illness, the great astronomer died at Slough while his son was traveling on the continent.

The famous house on the highway to Windsor, which

for many years had been the Mecca of astronomers, now became a house of mourning. In autumn of the same year Caroline Herschel, after half a century of devotion and self-sacrifice to her brother's work, returned to her native city of Hanover. John Herschel and his aging mother were left alone in the large house that had been so hospitable and frequented, and the young man became painfully aware of the weight of responsibility and of the difficulty of the tasks that now rested on his shoulders as the trustee of his father's intellectual heritage.

His scientific interests were still very diverse, although astronomy was now well to the fore, if only because of the collaboration with South. He nonetheless gave much of his attention to other scientific fields during this period. Time and again he became engrossed with problems of physics, and had he not had such a strong feeling of the need to preserve and extend his father's work, astronomy would even then have occupied only a peripheral position in his scientific activities. In 1823, he undertook with Babbage a series of experiments in electricity and magnetism, a field which even in the late eighteenth century had been an object of parlor games rather than of serious research except in the hands of such pioneers as the Italians Luigi Galvani (1737–1798) and Alessandro Volta (1747–1827), the American Benjamin Franklin (1706–1790), and others. Herschel communicated his paper "On certain motions produced in fluid conductors when transmitting the electric current" to the Royal Society in the form of the Bakerian Lecture for 1824.[20] In the following year he published in the *Edinburgh Journal of Science* a similar investigation, "On the mechanical effects produced when a conducting liquid is electrified in connexion with mercury."[21] Incidentally, William Herschel had also been much interested in the theory of electricity. Seven of his earliest papers, presented to the Philosophical Society of Bath, are on this subject, dealing mainly with the experiments that were customary at the time.[22]

John Herschel's researches on magnetism were related to a series of experiments with rapidly rotating metal disks which had been carried out by Arago. The rotation caused a magnetic needle freely suspended above the disk to be deviated from its initial direction and itself set into rotation. The experiments that Herschel carried out with Babbage in the latter's private laboratory in his house in London were described in the *Philosophical Transactions* for 1825.[23] These early experiments in the field of electromagnetic induction helped to lay the foundations for the invention of the shuttle winding for the dynamo by the German engineer Werner von Siemens (1816–1892) some thirty years later.

After 1820, Herschel did little research in chemistry; the literature shows only a short paper in the *Philosophical Transactions* for 1821, dealing with the chemical separation of iron from other metals.[24] Instead, his interest was turning to geology and mineralogy. His enthusiasm was greatly stimulated by a series of long journeys to various parts of Europe during the next few years. Apart from Clarke's London lectures, which he had listened to with great interest in 1816, he seems to have been led into these fields as a result of his polarization experiments which entailed handling a great variety of minerals. Some indicative remarks on the subject are to be found in a letter to Babbage in 1820, in which he writes of a planned geological tour of Ireland: "I shall eke out the time and try the temper of my new mineralogical hammer on the crags of Ilfracombe—or perhaps make a digression across the Bristol Channel into Glamorganshire—a county I have never visited. . . . Do you know of any geological work of moderate (i.e., portable) magnitude by which one can study the rocks by the wayside?" [25]

Herschel does not seem to have made the excursion to Ireland, but in the summer of 1821, accompanied by Babbage, he traveled through France to Switzerland and northern Italy. In Paris, where the two friends stayed for a week, Herschel called on Arago and communicated to him and other members of the Bureau des Longitudes some proposals from the Royal Society for a cooperative geodetic survey. He also met Laplace and the French physicist Jean Baptiste Biot (1774–1862), with whom he exchanged experimental results in physical optics. Biot, like Herschel, was carrying on experiments in polarization, and the two scientists were engaged in a lively correspondence. The German naturalist Alexander von Humboldt (1769–1859), who was living in Paris at the time, became acquainted with Herschel at Laplace's house.

On leaving Paris, Herschel and Babbage went southward through Dijon to French Switzerland, where they spent another week at Geneva. Herschel was greatly impressed with the active scientific life of that city, where he made the acquaintance of Professor Marc-August Pictet (1752–1825), president of the Geneva Academy of Sciences and one of a large family of learned men. The detailed description of the journey in Herschel's diary[26] throws an interesting light on the way scientists traveled in those days. The two friends took with them in their carriage a collection of the most varied scientific instruments. Wherever they went, they made barometric measurements of altitude and temperature measurements, determined angles with a small pocket sextant, and filled entire notebooks with their observations. The mountains of Switzerland afforded many opportunities for geological and mineralogical studies. Every peculiarity of terrain or populace was painstakingly observed and recorded. They visited historic cities, where Herschel made drawings with his camera lucida (see Chapter 4) to preserve memorable impressions in pictorial form. These sketches are supplemented by picturesque descriptions of moonlit nights and sunrises on lonely mountain tops, alternating with sober account of technical devices, records of scientific visits to astronomers and physicists, and anecdotal parentheses.

From Geneva they traveled southward through Chamonix, Aiguebelle, and Modane to Turin, where they visited the Italian astronomer Giovanni Antonio Amedeo Plana (1781–1864), and then to Milan. The return journey was by way of Lake Como, Lake Maggiore, and the Simplon Pass. A great mountaineering achievement was the ascent of the Breithorn, a formidable 13,000-foot height next to Monte Rosa, under the supervision of a Swiss mountain guide—a rash enough adventure for the two friends, who had no mountaineering experience.

Their journey continued through the Nicolai and Rhone valleys to Interlaken and Berne. At the famous Staubbach waterfall near Lauterbrunnen on the northern edge of the Berne Alps, Herschel made barometric measurements which he and Babbage reported in the *Edinburgh Philosophical Journal.*[27]

In the summer of 1822 Herschel toured Holland and Belgium in the company of his old friend the historian James Grahame. During this journey, which he recorded in detail in his diary, he received the news of his father's death.

A third major European trip, on which he embarked at the beginning of April, 1824, took him once more to Paris and to the south of France and through Italy down to the southernmost part of Sicily. Again he was hospitably received by scientific friends in all the larger cities. In Paris he renewed his friendly relations with Humboldt and Laplace and made new acquaintances among the most distinguished scientists of France. To his mother he wrote: "Thus within eight hours after my arrival in Paris, I found myself in company with almost everyone I wished to see: Arago, M. Laplace, . . . Baron Humboldt, . . . M. M. Thénard and Gay Lussac the chemists, Poisson and Fourier the mathematicians* and many more. . . . I breakfasted

* Louis Jacques Thénard (1777–1857); Siméon Denis Poisson (1781–1840); Jean Baptiste Joseph Fourier (1768–1830).

next morning with the Aragos at the observatory, where I found M. M. Bouvard and Nicollet,° and made all the observations I came for. We were joined by Humboldt who gave me some very useful information respecting the Alps, Vesuvius, etc." [28]

After a long and somewhat difficult journey through Savoy and over the snow-covered pass of Mont Cenis, with some involuntary delays caused by broken carriage wheels, Herschel reached Turin in mid-April. After a visit with Plana and a fine, clear observing night at his observatory, he continued to Genoa where he spent a few days as the guest of Baron Franz Xaver von Zach (1754–1832), an astronomer and the editor of the *Monatliche Corre-spondenz*, which was the earliest regular journal devoted to astronomy. In Modena he paid a visit to the Italian astronomer and optician, Giovanni Battista Amici (1786–1863), then went on to Bologna and Florence. "It is hardly possible to imagine anything finer than the vale of the Arno as seen from the height immediately above Florence, not to speak of the city itself, with its domes and towers and palaces. . . ." [29] In the middle of May he reached Rome, but the few days that he spent there provided only a fleeting inspection of the great city and its artistic treasures. At Naples he was cordially received at the house of the British ambassador. The name of Herschel seemed to work wonders, and all doors were opened to him. "In fact," he wrote to his aunt, "I find myself received wherever I go by all men of science, for his [William Herschel's] sake, with open arms, and I find introductions perfectly unnecessary." [30]

The ascent of Vesuvius was a climax in Herschel's itinerary. He measured altitudes with his barometer, made chemical analyses of the sulphurous vapors rising from the crater, and returned to Naples laden with rock speci-

° Presumably the astronomers Alexis Bouvart (1767–1843) and Jean-Nicolas Nicollet (1786–1843).

mens. Another day was given to viewing the excavations at Pompeii and Herculaneum. Herschel was fascinated, above all, by the amphitheater of Herculaneum, which at that time had been only partially dug out from under the masses of lava and ashes: "You stand on the stage . . . but it is 50 feet underground, and in the place of the applause of a cheerful audience, a black immovable rock like a huge gloomy curtain fronts you, and overhead the faint roll of the carriages in the streets of Portici murmurs in your ears like low thunder." [31]

On June 19 Herschel embarked from Naples harbor on a steamer which took him along the coast of southern Italy to Palermo. There he hastened to the sickbed of his father's faithful old friend, Giuseppi Piazzi (1746–1826), "who is a fine respectable old man, though I am afraid not much longer for this world," as Herschel informed his aunt in Hanover.[32] A member of the Theatine religious order, Piazzi had been a professor of mathematics and astronomy since 1781 and director of the Observatory of Sicily since 1817. His most important work was a large star catalogue containing 7500 objects. On New Year's Day, 1801, he had discovered Ceres, the first of the asteroids or minor planets of which the solar system contains thousands, mostly moving between the orbits of Mars and Jupiter. He had been one of William Herschel's most enthusiastic correspondents and had once visited the observatory at Slough.

From Palermo, Herschel started on a circular tour of Sicily. He visited the famous sulphur mines of Catolica, stood in front of the ruins of the temple of Agrigento, and ascended Mount Etna, spending a few days in Catania. The return journey began with a sea voyage back to Naples and continued by carriage to Rome and past Tivoli and the lovely Campagna by night to Perugia and Florence.

Instead of going directly across France, Herschel wished

to end his tour with a visit to Germany, his father's home-
land. He proceeded northeastward from Florence through
Padua to Venice, from which he wrote to his mother,
"The silence in the town from the total absence of carriages and horses is funereal, and the hearse-like appearance of the gondolas adds much to the melancholy of the
place." [33] He then traveled rapidly northward through
Verona and along Lake Garda to Trento, Bolzano, and
through the Brenner Pass to Innsbruck. He spent a few
days in the Dolomites, where he made a number of geological sorties and studied the language and customs of
the inhabitants. In Munich he visited the Bavarian optician and physicist Joseph von Fraunhofer (1787–1826)
in his laboratory and listened with excitement while
Fraunhofer explained the production of his achromatic
lenses and demonstrated a series of interesting optical
experiments. He presented Herschel with a large prism
of flint glass, which Herschel used twenty years later in
the course of his experiments in photochemistry, the
science of the chemical effects produced by light.

At Erlangen he sought out the German mathematician
Johann Friedrich Pfaff (1765–1825), who was editing a
German edition of William Herschel's works.[34] At Gotha
he met the German astronomer Johann Franz Encke
(1791–1865). Encke showed him through the telescope the
barely visible comet that carries his name, which Herschel
had already observed in Italy. At Göttingen University
Herschel visited Professor Karl Ludwig Harding (1765–
1834), the brilliant pupil and assistant of the deceased
German astronomer Johann Hieronymus Schröter (1745–
1816). The long journey was concluded by a visit to his
aunt Caroline at Hanover, and on the evening of October
18 John Herschel trod on English soil after an absence of
six and a half months.

Soon afterward he was again deep in his work. "Oct. 27.
Began an essay on physical optics." [35] This laconic note

in his diary marks the beginning of a major article that attracted much attention when it appeared three years later in the *Encyclopaedia Metropolitana* (see Chapter 3). In this article Herschel explicitly adopted the wave theory of light,* thus substantially contributing to the displacement of Newton's emission theory by the new ideas being advocated by Biot, Thomas Young, and Fresnel, the basis of which went back to the Dutch mathematician Christiaan Huyghens (1629–1695).

A further important event in Herschel's scientific life took place during 1824. In November, he was elected to the office of secretary of the Royal Society. That he should have received this honor suggests that the bad feeling between the Royal Society and the Astronomical Society had largely disappeared after the death of Sir Joseph Banks. Herschel found an understanding friend and supporter in the person of the new president of the Royal Society, the chemist Sir Humphry Davy.

The many duties of Herschel's post, particularly the necessity for regularly attending Royal Society meetings, induced him to move to London for the three years of his term of office, and he took up residence at 56 Devonshire Street, Portland Place.

Although these activities kept him in London for the major part of the year, he did manage a journey of several weeks to the continent in the autumn of 1826. His destination was the south of France and the Auvergne. In Paris he met the French naturalist Baron Georges Cuvier (1769–1832), who gave him useful information on geological

---

* The so-called undulatory or wave theory of light supposes that light rays are propagated in the form of waves going out in all directions from a luminous body. The medium in which the waves were propagated was considered to be a hypothetical extremely rarefied and completely transparent medium filling the whole of space called the "aether." Following later work by J. Clerk Maxwell, it was realized that the waves in question are electromagnetic in nature and the "aether" has not been regarded as a real physical medium since the acceptance of the theory of relativity.

topics and introductions to friends further along the route. With his faithful servant James Child, who had previously accompanied him on his long tour of Italy, Herschel traveled in bumpy mail coaches over the dusty highways of France, following the Loire and Allier valleys to Clermont-Ferrand and Montpellier. He ascended many of the volcanic domes and peaks of the mountainous part of the country, examined the geological strata, and visited the famous paleontological discovery sites near Issoire. "Upwards of 60 hitherto unknown animals (now extinct) have been found here," he reports in a letter home.[36] This is the first appearance of his interest in paleontology. On the Puy de Dôme he carried out, in addition to the usual barometric determinations of altitude, measurements "with a new instrument which I made before I left England and brought with me . . ."[37]

The instrument referred to, which Herschel called an "actinometer" or radiation meter, consisted of a device equipped with a sensitive thermometer designed to determine the flux of solar radiation on an absorbing surface of known area. It was a precursor of the pyrheliometer invented in 1837 by the French physicist Claude Servais Mathias Pouillet (1791–1868) and used in the determination of the solar constant.* Since the intensity of solar radiation is markedly greater at high altitudes than at sea level, because it suffers less absorption, Herschel preferred to carry out his measurements on high mountain peaks which protrude above the fog of the lower layers of the atmosphere. During his stay at the Cape of Good Hope, described in Chapter 4, he carried out several series of measurements with the actinometer, which were not appreciated at their full value until the growth of physical studies of the sun.

---

* The solar constant is defined as the amount of heat received from the sun in one minute by an area of 1 square cm. on the earth's surface placed at right angles to the direction of the sun, corrected for absorption by the atmosphere.

Herschel's last major journey before his expedition to
South Africa in 1834 was a tour of Ireland in the autumn
of 1827, on which Babbage accompanied him. He does
not seem to have had any particular scientific plans in
connection with this trip. At Dublin he visited the Observatory. Sir William Rowan Hamilton (1805–1865),
Royal Astronomer of Ireland and director of the Observatory, who was away at the time, expressed his regret at
having missed Herschel, in a warm letter: "We shall
always regard one another with friendly feelings, if you
will permit one so young and untried as myself to call
myself the friend of Herschel." [38]

Hamilton had been appointed Professor of Astronomy
at Dublin University and Royal Astronomer of Ireland in
1827 at the early age of twenty-two, and in 1837 he was
to become president of the Royal Irish Academy. His
greatest contributions to science were in the fields of
mathematics and theoretical physics. He was the inventor
of quaternions, a particular type of complex numbers used
in many kinds of mathematical problems, especially those
of physics, and the discoverer of conical refraction of light
rays during their propagation through biaxial crystals. He
was thus active in two fields in which Herschel was also
intensely interested, and these common concerns led to a
warm friendship and a lively correspondence which continued until Hamilton's death.

Herschel's visit to Ireland concluded his period of study
and travel. His scientific investigations, particularly in
astronomy, partly overlapped this period and continued
beyond it. The immense variety of his interests, already
displayed, is at least equally evident in the years before
the expedition to the Cape of Good Hope, which constituted one of the most creative phases of his life.

3

The Astronomical Heritage

✦ ✦
✦

In the period from 1825 to
1833, the middle years of John Herschel's life, his most
important publications were two great catalogues. The
first is a list of some 2300 nebulae and star clusters,[1] the
result of a complete revision and extension of William
Herschel's nebular observations. The second is a monu-
mental six-part catalogue of double stars [2] which bears
witness to extraordinary diligence in observing and minute
attention to detail; these qualities are the ones most essential in this field, where repeated measurements of the
utmost precision are required for each individual system.

Herschel was predestined to excel in this field of astronomy both by training and the excellent instruments
he had at his disposal. He had acquired from South the
7-foot equatorial which they had used together. It was
ideally suited to double-star work and Herschel was accustomed to its use from long practice. For nebulae and
clusters, the observation of which required chiefly a telescope of great light-gathering power, he got excellent
service from the 20-foot reflector, whose optical parts he
and his father had constructed; of its three interchangeable mirrors, he had produced one with his own hands.
The "twenty-footer" was the largest telescope in the world
at that time. In a letter to Caroline Herschel dated April

18, 1825, he announced his plan to pay special attention to nebulae. "These curious objects . . . I shall now take into my especial charge—nobody else can see them." [3]

The task he had set himself was by no means an easy one. Caroline's Zone Catalogue of the nebulae and clusters observed by William Herschel [4]—a remarkable work produced by that indefatigable astronomer in her old age—was most helpful in this enterprise, but John was not in his father's fortunate position of having a faithful assistant by his side for dozens of years. He had to rely exclusively on his own efforts. For the purely manual task of setting and guiding his heavy, unwieldy telescope, he procured the services of a mechanic, John Stone, who later accompanied him on his expedition to South Africa.

He swept the entire northern hemisphere of the sky in the same manner as his father had done. His catalogue of the nebulae differs from his father's in ordering each object according to right ascension and in giving the north polar distance * for every nebula and cluster. All the coordinates were referred to the equinox of 1830,† which made the catalogue easy to use and ensured that the data were on a reasonably uniform system. The 2306 nebulae and clusters in the catalogue included 525 newly discovered objects; this was a respectable supplement to William Herschel's discoveries, considering how thoroughly the latter had repeatedly searched the heavens. Thus John

---

* Polar distance is the angular distance of a celestial object from the north or south celestial pole. Nowadays, instead of polar distance, astronomers generally use declination—that is, the angular distance from the equator, counted positive to the north and negative to the south. Declination corresponds to geographic latitude on the earth's surface.

† Owing to certain gravitational disturbances, the earth's axis undergoes a toplike motion in space (precession). This leads to a steady drift of the equinoxes (intersections of the celestial equator and ecliptic) and to a secular or systematic change in celestial coordinates (right ascension and declination). Consequently all measured positions have to be corrected, or "reduced," to a particular epoch, the equinox for which is taken as the origin of the coordinate system, before they can be incorporated into an astronomical catalogue or chart.

Herschel remarks in the introduction to his Catalogue: "It may serve to show the close and rigorous nature of my father's scrutiny, when I state, that among these 500 I can call to mind only one very conspicuous and large nebula, and only a very few entitled to rank in his first class, or among the 'bright nebulae.' By far the greater proportion of them are objects of the last degree of faintness, only to be seen with much attention and in good states of the atmosphere and instrument." [5]

Herschel describes his method of working in a letter to his aunt: "But I always find yr catalogue most useful. I always draw out from it a regular *working list* for the night's sweep, and by that means have often been able to take as many as thirty or forty nebulae in a sweep. I have now secured such a degree of precision in taking the places of objects in the telescope that the setting stars . . . cross the wire often on the very beat of the chronometer when they were expected . . . In short, I reckon my average error in R.A. in determining the place of a new object by a single observation, not to exceed one second of time, and in Polar distance a quarter of a minute. This you will easily perceive to be a considerable improvement in respect of precision, which is more my aim than it was my father's, whose object was only discovery. I have found a great many nebulae not in your catalogue, and which, therefore, I suppose are new . . ." [6]

The unadorned figures and tables of Herschel's catalogue conceal an extraordinary amount of painstaking, detailed work. Observing conditions for star clusters and nebulae in the English sky were far from ideal. Absolutely clear nights with an atmosphere completely free from dust and mist are quite uncommon in England, and only moonless nights are of any use for nebular observations. In addition those regions of the northern sky that are most endowed with nebulae and therefore produce the greatest observational yield are confined to a relatively small region in the constellation of Virgo, which can be observed in the first half of the night only in the spring months. In a letter to Francis Baily, Herschel gives a glimpse of the difficulties with which he had to contend. "Every hour is precious," he writes, "from the circumstance of the great mass of nebulae lying in the 11, 12 and 13 hours of R.A., and which therefore must be observed in the spring or not at all. A pellucid sky—the total absence of moonshine and twilight—and *nebulae to look at*, are conditions which coincide, on the average, not 20 nights in the year, and the sacrifice of a night . . . is therefore a very serious evil to me, regarding as I do the completion of this work not as a matter of choice or taste, but a sacred duty which I cannot postpone to any consideration." [7]

Herschel's catalogue of nebulae made no claim to completeness, nor did it purport to give an exhaustive description of the various objects observed. The completion of such a task would have filled more than one lifetime. The immediate purposes of the enterprise were to provide an enduring monument to William Herschel's work and to give observing astronomers a ready reference to enable them to find particular objects. An additional objective was to enable comet observers to decide definitely whether a newly found nebulous object was a comet or one of the many nebulae and clusters that can easily be confused with comets before they reach perihelion (the point of orbit nearest the sun). A similar task had been undertaken half a century earlier by the French astronomer Charles Messier (1730–1817). Messier's catalogue,[8] however, only contained 100 or so of the brightest and most conspicuous nebulosities; the number in Herschel's provides an impressive demonstration of the progress of telescope technique in the course of those fifty years.

Herschel's catalogue also contains nearly 100 drawings of the most remarkable objects. These small sketches were executed with great care and are among the most important graphic illustrations of celestial objects made before the introduction of photography. Although they do not compare well with modern photographs, it would be wrong merely to dismiss them on the grounds of the superiority of the present-day photographic plate. The essential details are recorded correctly. Furthermore, the visual impression caused by a nebula or a star cluster even in the best telescopes is always different from its appearance in a photograph, partly because the human eye and the photographic plate have different relative sensitivities in different regions of the spectrum, and partly because, unlike the eye, the plate can add up the light reaching it over very long periods of time and is thus able to record much fainter signals.

In 1826 Herschel published a monograph on the Orion nebula and one on the nebula in Andromeda, together with other observations made with the 20-foot telescope.[9] The purpose of the paper on the Orion nebula was to determine from a comparison with earlier observations whether any noticeable changes in structure or brightness had occurred. The text is supplemented by a drawing of the nebula.

During the same period he was preparing the catalogue of double stars already mentioned. This was similar in arrangement to and no less important than the catalogue of nebulae. Here again John Herschel showed himself a worthy successor to his father; in fact, considering the 5075 objects listed in his catalogues compared with about 850 discovered by his father, John Herschel can be claimed to have established double-star astronomy in England. Although a number of these objects would not be regarded nowadays as true physical pairs, because of the wide separation of the components, Herschel's catalogue is nevertheless one of the earliest reference works of its kind to be comprehensive and of practical use. While a substantial number of double stars were taken from the lists prepared by the Herschels' only serious competitor in this field, Wilhelm Struve, John Herschel himself discovered 3347 systems. Herschel's greatest contribution lay in bringing the enormous amount of observational material together, checking the data for double stars discovered by others with his own observations, and giving a consistent list arranged in order of right ascension for the equinox of 1830.

This mighty work was completed single-handed in barely ten years of intensive observing. Naturally the enterprise had its share of disappointments and phases of deep discouragement. "Two stars last night, and sat up till two waiting for them," Herschel once wrote. "Ditto the night before. Sick of star-gazing—mean to break the telescope and melt the mirrors." [10] However, these occasional bouts of depression were usually of short duration.

Stimulated by his innumerable measurements of binaries, Herschel attempted to develop a general method for the determination of double-star orbits.[11] Starting from his experience that measurements of separation are affected by considerable uncertainties, if only because there is often a large difference in brightness between the two components, he based his method exclusively on measurements of position angle and on the time intervals between separate observations. Unlike the other methods current in his time, Herschel's is based entirely on graphical construction. The observed values of position angle and time intervals are plotted in a coordinate system to yield points on a curve from which is derived the observed orbit of the companion projected on the sky. Assuming that double-star orbits are actually ellipses with the primary at one focus, in accordance with Kepler's first law, Herschel was able to use an averaged interpolation curve through the observations to compute the true elliptical orbit described by the star in space.

For this work he was awarded a Gold Medal by the Royal Society on November 30, 1833. Although his graphical method has long since been superseded by other procedures based on improved techniques of observation and measurement, it was of historical significance as one of the first applications of computational astronomy to the fixed stars. The Duke of Sussex (Augustus Frederick, 1773–1843), president of the Royal Society, gave the following appreciation of Herschel's achievement in an address to the Fellows: "Herschel has observed and registered many thousand distances and angles of position of double stars, and has shown, from the comparison of his own with other observations, that many of them form systems, whose variations of position are subject to invariable laws. He has succeeded, by a happy combination of graphical construction with numerical calculations, in determining the relative elements of the orbits which some of them describe round each other, and in forming tables of their motions; and he has thus demonstrated that the laws of gravitation which are exhibited as it were in miniature in our own planetary system, prevail also in the most distant regions of space; a memorable conclusion justly entitled by the generality of its character to be considered as forming an epoch in the history of astronomy, and presenting one of the most magnificent examples of the simplicity and universality of those fundamental laws of nature, by which their great Author has shown that He is the same today and for ever, here and everywhere." [12]

Herschel's preoccupation with double stars, his guiding passion between 1825 and 1833, led also to another piece of research. He worked out a method of deriving the so-called annual parallax, and hence the distance, of a double star from the variation in its position angle. The term "annual parallax" refers to the change in direction on the imaginary surface of the celestial sphere that a star undergoes when seen from two diametrically opposite points of the earth's orbit around the sun. Owing to the enormous distance of stars relative to the baseline, this angular shift is very minute. It is found by measuring the position of the star one is interested in as accurately as possible in relation to more remote stars in the neighboring portion of the sky and comparing the results obtained at intervals of six months—that is—half the period of the earth's revolution. Since the earth travels from one end of a diameter of its orbit to the other during this interval, a star that is sufficiently close to it will undergo a slight change in its position relative to another star that is farther away. It was this consideration that originally drew William Herschel's attention to double stars. He had considered them to be especially suitable for parallax measurements, assuming that their close proximity in the sky (at least in the case of pairs having unequal brightness) was due to chance. However, his repeated observations and measurements proved that most of these binary pairs were real physical systems held together by gravitation, so that they could not possibly be used for direct determinations of parallax. Even in the case of the pairs whose two components are indeed seen together as a result of chance and are in reality at quite different distances from us, he could not detect any change in apparent separation, simply because his micrometer was not sensitive enough to measure minute angles of a fraction of a second of arc.

John Herschel in his method again made use of these so-called optical (apparent) pairs, but he measured the position angle instead of the separation. In his paper "On the parallax of the fixed stars" (1826),[13] he wrote: "I do not find that it has been noticed, however, that parallax must occasion a periodical change in the angle of position, as well as in the distance of the two stars composing a double star, and that this variation is much more conceptible of ready and exact appreciation with our present micrometers than that of their distance." [14]

The parallactic shift undergone by a star as a result of

the earth's revolution round the sun causes it to describe a small ellipse in the sky—a tiny mirror-image of the earth's annual orbit. The nearer the star is to the earth, the larger is this "parallactic ellipse." If, then, one is looking at an optical pair with the two members at very different distances, the remoter component will have no or hardly any perceptible parallax. Herschel constructed the following example: "To estimate the extent of this variation, let me conceive two stars so situated as to have their apparent line of junction in the direction of a secondary [perpendicular] to the ecliptic, and therefore at right angles to the major axes of their parallactic ellipses— let their distances from us be such that the nearer one shall have a parallax of 1″, and the farther one no appreciable amount of it. Also, let their apparent angular distance from each other be 5″. It is evident that the variation alluded to will equal the angle subtended by a line of 1″ in length, at a point 5″ distant from its middle, that is, to 11° 25′. Now this is a quantity which is quite beyond all conceivable limits of error of observation in the measurement of double stars, and for stars nearer than 5″ the amount is of course proportionally greater. Thus for two stars, at only 1″ distance from each other, of which the one is affected by parallax to the amount of 1″, and the other not at all, the annual variation in position will amount to upwards of 53°." [15]

In a table at the end of his paper, Herschel gives annual parallaxes for some seventy stars inferred from their variations in position angle. The values range from 0.013″ to 0.136″ and agree quite well in order of magnitude with modern values. Although they cannot be said to be the first definitive determinations of parallax—which were actually achieved about ten years later by the Scottish astronomer Thomas Henderson (1798–1844) and by Bessel [16]—they were nevertheless an important step toward the solution of this problem. Up till then, it had

been possible to say with certainty only that annual parallaxes must be smaller than a second of arc. Now for the
first time someone had tried to estimate their actual values.

Herschel himself was characteristically hesitant and
modest in assessing the importance of what he had done.
On December 30, 1825, he wrote to his aunt in Hanover:
"Do not suppose that I pretend to have discovered parallax, but if it exists to sensible amount, I think it cannot
long remain undiscovered if anybody can be found to put
into execution the method I am about to propose; and I
hope it will be taken up by astronomers in general." [17]

In February 1827, Herschel was elected president of
the Astronomical Society. The office was largely an honorary one, involving the duties of presiding over the monthly
meetings of Fellows, representing the Society at official
functions, receiving distinguished guests and carrying out
other representative duties, but Herschel found that it
seriously interfered with his research and literary activities. The honor of being chosen for the presidency and
the prestige that went with the office were not enough to
make up for this. He wanted above all to keep his personal freedom and scientific independence, and not to
have the scope of his private researches restricted by any
official and public engagements. He wrote a friend early
in 1828: "It would be my most earnest wish, I confess, to
be allowed to retire from the office of Pres. [of the Astronomical Society] at this anniversary." [18]

For the same reason he had resigned his secretaryship
of the Royal Society in December 1827. He had written
to Wilhelm Struve on January 3, 1827: "I intend, after the
present session of the Royal Society is terminated, to
resign my office of Secretary, and I hope then to have
more time to devote to astronomy than hitherto." [19]

He also refused two very tempting offers that were
made to him during this period. One was an urgent and
well-meant suggestion by Whewell and other Cambridge

friends that he should apply for the vacant Lucasian Professorship, the Chair once occupied by Newton. The other was an offer from Lord Brougham, one of the founders of the University of London, of a Chair of Mathematics in that recently established institution. Herschel wrote to Brougham: "To teaching I have a positive dislike and would certainly engage in no office in which it formed any part of the duties expected from me." [20] To Whewell he admitted, while proposing their mutual friend Babbage for the Lucasian Chair, that it was partly out of vanity that he would prefer his contribution to knowledge to be regarded as that of an amateur rather than a professional scientist. He wrote: ". . . possibly too [it is] a kind of obscure consciousness that I am not destined . . . to make giant inroads into great branches of human knowledge— but rather to loiter on the shores of the ocean of sciences and pick up such shells and pebbles as take my fancy for the pleasure of arranging them and seeing them look pretty." [21]

This statement, which paraphrases a similar declaration by Newton, is typical of John Herschel's character and of his attitude to science. At first sight, such an apparent excess of modesty comes strangely from a man who had already made a brilliant name for himself in the scientific world at the time when he wrote this passage. It would probably be a mistake, however, to regard Herschel's frequent statements in this vein as merely a sign of his modesty. Equally important in them is a desire to follow his various scientific interests independently and in peace, without being committed to any particular direction or specialized target, so that he could be aware of the seductive breadth of the "ocean of sciences," but at the same time not overlook the beauty and significance of any particular shells and pebbles that might be washed ashore by the waves. An essential factor in Herschel's scientific drive was an aesthetic appreciation of the inner beauty of things, combined with pleasure in the advancement of knowledge for its own sake. At the same time, he did not restrict himself to the playful outlook expressed by his metaphor of shells and pebbles but searched continually for the logical connection between isolated facts with the aim of establishing general underlying laws. In short, he never lost sight of the ocean itself.

"The moment we contemplate nature as it is, and attain a position from which we can take a commanding view, though but of a small part of its plan, we never fail to recognize that sublime simplicity on which the mind rests satisfied that it has attained the truth," Herschel wrote at the end of a book entitled *Preliminary Discourse on the Study of Natural Philosophy*.[22] This work, first published as the opening volume of the *Cabinet Cyclopaedia*, edited by the Irish physicist and mathematician Dionysius Lardner (1793–1859), appeared separately in London in 1830 and was translated into French, German, and Italian. A second edition appeared in 1851. The term "natural philosophy" in the title was used in the sense of "natural science," as is expressed, for example, by the title of the German version, *Über das Studium der Naturwissenschaft*. In Germany "*Naturphilosophie*," originally embracing every kind of study of nature, had been restricted since the time of the philosopher Immanuel Kant (1724–1804) to the sense of a purely speculative philosophy, whereas in England the corresponding term continued to be used in its classical connotation throughout the nineteenth century and was not replaced by the rather colorless word "science" until the twentieth.

Herschel's book appears to have been highly successful in the best sense of the term. The great naturalist Charles Darwin (1809–1882) acknowledged that he had received the greatest inspiration for his own researches from Herschel's *Preliminary Discourse* and Humboldt's *Reise in die Aequinoktial-länder der Neuen Welt*.[23]

The *Preliminary Discourse* provides both a general introduction to the nature and method of science and short surveys of its various branches, although the subjects described are actually restricted to the "physical" sciences —astronomy, physics, chemistry, geology, and mineralogy. The main weight of the treatment lies in the section on method, which is much more exhaustive than might be expected in a mere introduction. Apart from historical remarks, there is an extensive discussion of purely philosophical considerations on the meaning and utility of science, as is indicated by the very table of contents of the first chapter: "Of man regarded as a creature of instinct, of reason, and speculation.—General influence of scientific pursuits on the mind." However, Herschel was not concerned with *Naturphilosophie* in the Kantian sense. He did not treat natural phenomena as an object of philosophical speculation, but used philosophy as an aid to the understanding and generalization of the "exact" results of observation and experiment.

Despite its flowery and intricate style, Herschel's book still makes stimulating reading today because of the immense enthusiasm and delight with which the author discusses his subject. Even present-day readers, under the stress of modern life, will appreciate the following words: "It is not one of the least advantages of these pursuits, which however they possess in common with every class of intellectual pleasures, that they are all together independent of external circumstances, and are to be enjoyed in every situation in which a man can be placed in life. . . . They may be enjoyed, too, in the intervals of the most active business; and the calm and dispassionate interest with which they fill the mind renders them a most delightful retreat from agitations and dissensions of the world, and from the conflict of passions, prejudice, and interests in which the man of business finds himself involved." [24]

The *Preliminary Discourse* is devoted mainly to educational and ethical ends, rather than to specialized scientific training. It is intended to inspire the reader to think about scientific problems and to draw the attention of people driven by the turmoil of business affairs and by preoccupation with material needs to the existence within this material world of a spiritual sphere whose delights "are to be enjoyed in every situation in which a man can be placed in life." Herschel did not aim in his book to give a summary of the state of science in his time, nor to provide a compendium or encyclopedia, but to point out the spiritual and moral foundations of scientific research. Seen in this light, his book is to a great extent a philosophical work. Much of its historical importance, also, lies in its deliberate departure from the materialistic point of view dominant at the time, partly under the influence of the philosophy of the Enlightenment and partly owing to the impression created by the astounding progress of science. Thus Herschel writes that the very advance of science arouses man's consciousness of the limitations of his actual and possible knowledge and shows him "a distant glimpse of boundless realms beyond." [25] "Is it wonderful," he continued, "that a being so constituted should first encourage a hope, and by degrees acknowledge an assurance, that his intellectual existence will not terminate with the dissolution of his corporeal frame, but rather that in a future state of being, disencumbered of a thousand obstructions which his present situation throws in his way, endowed with acuter senses, and higher faculties, he shall drink deep at the fountain of beneficent wisdom for which the slight taste obtained on earth has given him so keen a relish?" [26]

Herschel also expresses his own attitude toward the utilitarian approach, denying all spiritual values, based on the philosophy of materialism and greatly strengthened in Herschel's own time by the rapid progress of the Industrial Revolution: "The question 'cui bono'—to what practical end and advantage do your researches tend?"—is one which the speculative philosopher who loves knowledge for its own sake, and enjoys, as a rational being should enjoy, the mere contemplation of harmonious and mutually dependent truths, can seldom hear without a sense of humiliation. He feels that there is a lofty and disinterested pleasure in his speculations which ought to exempt them from such questioning."[27] These words were a resounding challenge to the negative scientific outlook of the age, which was tending to degenerate from jejune rationalism to a still more barren materialism.

The *Cabinet Cyclopaedia*, of which Herschel's *Preliminary Discourse* was the introductory volume, was one of many encyclopedias then beginning to appear all over Europe after the pattern of the great French *Encyclopédie* of Denis Diderot and his colleagues in the mid-eighteenth century. Most of these were many-volume works which gave surveys of the various fields of the arts and sciences in more or less extensive monographs. Herschel also contributed to Lardner's encyclopedia *A Treatise on Astronomy*, an introduction to astronomy designed for the general reader, produced at about the same time as the *Preliminary Discourse*. It was published separately in 1833,[28] and later editions appeared up to 1845. In 1849, Herschel republished it in greatly expanded form as *Outlines of Astronomy*. This famous book is discussed in detail in Chapter 7.

A much more important work of Herschel's is the long treatise on "Light," which he wrote in 1827 for the *Encyclopaedia Metropolitana*,[29] a similar undertaking to the *Cabinet Cyclopaedia*. It was translated into French in 1830 and into German in 1831. The beginnings of the work go back to 1824. Even a cursory examination of this impressive article, which fills 245 quarto pages, reveals its uncompromisingly scientific presentation, in marked contrast to the style of the *Treatise on Astronomy*. Even the latter is hardly a "popular" work in the sense in which the term is used today, but it is readily accessible to the educated layman familiar with very elementary mathematics and physics. "Light" is a dialogue between expert and expert, couched in the rigorous language of mathematical and physical proofs and formulas, interspersed with terse descriptions of innumerable optical experiments. It makes rather dry reading, understandable only by mathematicians and physicists, and this was the author's intention. Even a nonspecialist can see that here Herschel is in his native element. Had he not once declared that light was his first love?

Apart from giving a complete survey of the field, the treatise contains a great deal of highly original material, including the numerous discoveries, especially in the field of polarization, which Herschel had made in the course of his experiments with various minerals. Many of these experiments are elegant and subtle, giving a lively impression of the beauty and variety of optical effects. In later years, Herschel often used to demonstrate these experiments to his family and friends. Using simple means, such as a candle and a few crystals, prisms, or pieces of colored glass, he brought the most intricate optical phenomena to life in front of his audience and made complicated and difficult physical laws easy to understand. The article in the *Encyclopaedia Metropolitana* is not a mere survey, but actually a textbook on optics. He himself, however, does not seem to have regarded it as his *magnum opus* in this field. In January 1829, he wrote to Caroline Herschel at Hanover: "This work has excited a much greater sensation than I expected it would. . . . It is now translated into French and German and will thus I fear anticipate the effect of a great work I mean to write on the same subject." [30]

The "great work" he alluded to was never written, but

its effect could hardly have been greater than that of the encyclopedia article, which did much to prepare the way for the adoption of the wave theory of light, up to then not widely accepted in England. Herschel attempted to make the first comprehensive survey and appreciation of the investigations by Fresnel, Young, Brewster, Arago, Biot, and others, which had placed the wave theory on a firm scientific basis. Although his earlier publications had been based on Newton's ideas, he had now become a firm advocate of the wave theory and gave a detailed exposition of its advantages over Newton's hypothesis. At the same time he made no attempt to cover up the weaknesses and uncertainties still present in the new theory, and indeed in any theory of light then available. "The fact is," he writes, "that neither the corpuscular nor the undulatory, nor any other system which has yet been devised, will furnish that complete and satisfactory explanation of all the phenomena of light which is desirable. . . . The undulatory system especially is necessarily liable to considerable obscurities; as the doctrine of the propagation of motion through elastic media is one of the most abstruse and difficult branches of mathematical inquiry. . . ." [31]

As Herschel was well aware, the wave theory was not capable of accounting for all optical phenomena, but this weakness is common to all scientific theories. There is always a residue of effects and phenomena that do not conform to the requirements of the theory, or that do so only imperfectly. Consequently Herschel made no attempt to use his scientific eloquence and skill in advocacy of the wave theory to the exclusion of any other hypothesis, but advised the reader "to suspend his condemnation of the doctrine for what it *apparently* will not explain, till he has become acquainted with the immense variety and complication of the phenomena which it will. . . ." [32]

The article is divided into four main parts. The first

describes the properties of unpolarized light. Starting with photometry (the measurement of the intensity of light), he proceeds to explain the phenomena of geometrical optics, the laws of reflection and refraction, the structure of the eye, and the physiology of vision. This section concludes with a brief survey of the most useful optical instruments. The second part provides an introduction to the theory of color. There is a detailed account of the dispersion of light by prisms, a long chapter on the theory of the achromatic telescope, and a description of the absorption of light in noncrystalline substances and the production of colors. The third part confronts the two main theories of light and gives a detailed exposition of interference and diffraction in the framework of a general development of the wave theory. Both this chapter and the fourth part, which describes polarization effects, are completely inspired by the wave theory. Incidentally, in a systematic treatise a review of the existing theories would normally be placed either at the beginning or at the end. Herschel places it in the middle of his. This was certainly done deliberately, since he wished to emphasize that it was one of the aims of his work to help the wave theory to be accepted.

Herschel wrote another contribution to the *Encyclopaedia Metropolitana* in the field of physics: a short survey of "Sound." [33] This appears in the same volume as the article on "Light," although it was completed three years later (the "Light" article is dated December 1827, the "Sound" article February 1831). It is much shorter and contains no new results or descriptions of his own experiments; it is, in fact, just an article for an encyclopedia, well constructed and providing the essentials of this branch of physics.

Another enterprise of Herschel's, though not in the main stream of his research activities, was none the less informed by the same methods and shows him as a practical scientist. The object of this exercise was to determine the difference in geographic longitude between the Observatory at Paris and that at Greenwich by comparing observations taken from both. Similar measurements had already been carried out at the various observatories on the continent, but Greenwich had not been tied into the system. Progress in transport and communication had led to a rapid growth of cooperative international projects in astronomy, so that exact knowledge of the differences in longitude between the principal observatories had become a matter of considerable importance.

The enterprise was carried out by the British Board of Longitude in collaboration with a group of French scientists and engineers. The British group was led by Herschel and his friend Colonel Edward Sabine (1788–1883), an astronomer and surveyor particularly well qualified for the task. Sabine's knowledge of geodesy and practical experience in various expeditions made him an ideal member of the team.

The method of operation was to establish a chain of temporary observing stations between the Greenwich and Paris observatories. These stations had to be erected on outstanding points of the landscape so that they could communicate by means of visual signals (rockets). These light signals, which were to be sent out at regular prearranged intervals, were to be recorded, using chronometers of especially high accuracy. The observation of the signals was facilitated by the use of Dollond night-glasses of high light-gathering power.

Herschel had selected two observing sites in England: one was Wrotham, a small village some 20 miles southeast of Greenwich, situated on a hilltop overlooking a large area of the surrounding countryside, with particularly well-determined geographic coordinates because it had been used as a fundamental trigonometrical point in the Ordnance Survey of 1822. At an altitude of 770 feet above

sea level, Wrotham was the highest point between Greenwich and the coast. The second observing site was the
hamlet of Fairlight Down, near Hastings, which was also
a trigonometrical point in the survey. The observing station
erected here was situated above the steep slope of the
coastal cliff at an altitude of 578 feet above sea level, and
at a distance of some 34 miles in a direct line from
Wrotham. Across the English Channel the first site was
the French coastal station of La Canche at the mouth of
the river Canche. The next was the village of Lignières,
southwest of Amiens and 53 miles in a direct line from
La Canche, the longest distance between sites. The last
substation before Paris was Mont Javoul, about 40 miles
from both Lignières and Paris Observatory. The total
distance between the Greenwich and Paris observatories
was some 230 miles. The determination of the difference
in longitude between them was thus divided up into a
series of sections, each one of which could be readily
determined by observation of the light signals with the
aid of a chronometer. Addition of the individual differences gave the total.

The measurements were carried out in July 1825. Numerous observations gave a final mean value of 9 minutes
21.6 seconds of time (2° 20′ 24″ of arc) for the longitude
difference between Greenwich and Paris. In view of the
immense difficulties associated with the undertaking this
result is surprisingly accurate; the modern value of 9
minutes 20.9 seconds (2° 20′ 13.5″) is only 0.7 seconds of
time (10.5″ of arc) less than Herschel's.

Nowadays, such measurements are easily made with
radio signals that can be transmitted all over the earth
with the speed of light, and there is no need to divide up
large intervals into smaller ones with visual communication. The possibility of making such determinations with
the aid of the electric telegraph was pointed out in 1839
by the German mathematician Karl Friedrich Gauss

(1777–1855), but several decades had to pass before telegraphic communication finally replaced the older methods, including that of the transfer of chronometers which had become famous as a result of Struve's work. Herschel's report[34] describes how a whole detachment of artillery had to be detailed to fire the rockets and gunpowder signals at each of the various stations, and how a group of scientists worked intensively for twelve nights, observing, measuring, preparing tables, and computing. The immense elaboration of the operation shows what efforts were then needed to achieve results that are now everyday scientific practice, and also how a single discovery or invention can eliminate the need for enormous requirements in manpower and materials.

Apart from his functions as secretary of the Royal Society and of the Astronomical Society, Herschel had up to that time led the withdrawn life of a private scholar, but his selection as leader of this enterprise of international importance shows that he already had an established and distinguished position in the scientific life of the country. The very name Herschel commanded so much prestige, both in England and in the rest of Europe, that a bearer of it who had himself achieved as much as John Herschel had was bound to arouse the interest of the scientific public. Herschel would have avoided this if he could. He desired above all to carry out his researches in freedom and independence, untrammeled by any public office or other official duties. It was the tragedy of his scientific life, especially in later years, that he was continually weighed down by offices, official duties, commissions, and honors of all kinds.

At this period, however, he was still in the happy position of being the master of his time and plans and able to develop his various scientific interests without interference. His father had left him a very considerable fortune, which enabled him not only to fill all material needs but to finance his sometimes quite expensive scientific enterprises without requiring a professional salary. He had many friends and a large number of scientific correspondents in Great Britain and abroad. His life was filled by a multitude of scientific tasks which he set himself and brought to a successful conclusion with the obsessive energy inherited from his father. Above all, however, he had dedicated his full powers to completing his father's work in astronomy.

To carry out that task adequately without giving up his own intensive scientific interests required the sacrifice of many of the activities that enrich the lives of normal people. John Herschel's life was thus deprived of inner harmony and balance; he seldom relaxed or rested, nor did the sort of life he led enable him to draw strength from simple human contacts in addition to his intellectual resources. He had no easygoing social activities with congenial companions. Most of his friendships had arisen from shared scientific interests, and of these scientific friends only a few—Whewell, Grahame, Peacock, and Babbage— were intimates. The weekly dinners at the Royal Society Club were formal affairs, hardly suited to the forging of social contacts. Furthermore, the house at Slough had become very lonely since his father's death. His aging mother lived there in seclusion, and although Herschel was devoted to her, there was no cheerful domestic atmosphere to provide a focus of relaxation in his restless existence. He was constantly traveling between Slough, where he spent his nights at the telescope, and the house he had rented in London. His exhausting way of life did not agree with his constitution, not very strong at best, and occasionally he reached the verge of complete nervous collapse. The constant turmoil of his life was beginning to cramp his spirit; soon he was well on the way to degenerating into a crotchety and eccentric old scholar.

His friend James Grahame seems to have observed this

danger. Grahame had been happily married for some years and had secured through his family life the healthy balance that Herschel lacked. In a series of letters Grahame urged his friend to marry and raise a family as he himself had done, using all his eloquence to recommend this course as though it were a kind of medicine. Nor did he confine himself to suggestions and generalized recommendations. He had conceived a definite plan to introduce his friend, who was not exactly a ladies' man, to female society. He was acquainted with a Mrs. Alexander Stewart, the widow of a Scottish Presbyterian minister, who lived in London with her two daughters and six sons. Without further ado, he introduced Herschel into her hospitable household.

Grahame's extremely direct and simple plan achieved remarkably rapid success. Herschel soon felt at home at the Stewarts'. He was first attracted merely by their harmonious family life and natural cheerfulness, but soon Grahame observed to his satisfaction that Herschel was becoming very friendly with the younger daughter, Margaret, an unusually pretty and intelligent girl of eighteen. The friendship developed into an attraction, which in turn blossomed into passionate love. Herschel felt awakened to a completely new life that he could scarcely understand and that broke into his previous mode of existence with a sudden and violent burst, filling him with joy and confusion. Eager to confide his feelings to Grahame, Herschel followed him to Southampton, where Grahame, who was starting on a trip to France for his health, was awaiting the ferry. Before the two friends parted, they held a lengthy dialogue in a romantic setting, recorded in Herschel's diary: "Rowed by moonlight to Netley Abbey, where passed an hour in a close and whispered confidence in the main aisle under the ash trees, with the stars looking in above. An unearthly scene!" [35]

The object of Herschel's increasingly frequent visits to

the Stewarts could not be concealed for long. A letter from Grahame to the eldest son, Patrick, seems to have laid the groundwork for the decisive step. After describing Herschel's virtues in glowing terms, Grahame concluded: "If Margaret were my daughter I would say to her: 'My dear, you have gained the attachment of one of the best and most admirable men in England. If you can love him, I will venture to predict, as far as mutual foresight may, that you will be a happy woman.' What more can I add?" [36]

Mrs. Stewart was a lady with a strict regard for etiquette. Apparently, on receipt of Grahame's fervent testimonial she gave Herschel to understand that he was requested to send his proposal to her daughter in writing. This he did and on the next day received a very stiff reply from Margaret that evidently had passed her mother's censorship. It began with the customary formula: "It was indeed with feelings of the greatest surprise that I received your letter." [37] Perhaps Margaret's surprise was not really as great as she claimed, but in any case she accepted Herschel's proposal. The two were married at St. Marylebone Church, London, on March 3, 1829.

At the time of his marriage, Herschel was a mature man about to embark on the second half of his life, while his bride, nearly twenty years his junior, was still only a young girl. Nonetheless the marriage proved highly successful, "a union of unclouded happiness," says Herschel's first biographer, Agnes M. Clerke.[38] Margaret had had the benefit of an excellent education, and with all her youthful exuberance she was nonetheless very mature for her age. She was well aware of what she was doing in joining her life to that of a serious-minded, reticent scholar wrapped up in his own ideas. She was prepared to stand beside him and share his interests, to smooth his path and to enter into his world, strange though it was to her at first, at the expense of many of her own personal wishes and expectations.

Herschel realized that he would have to introduce his young wife to his own special world with considerable care. After a honeymoon of several weeks in the west of England, he decided to take her on a tour of the Continent, especially of France, Germany, and Italy. All the details of this trip are preserved in his diary.[39] On returning to England, the couple took up residence in his house in Devonshire Street, and Herschel, in addition to his normal scientific and literary activities, began planning an enterprise on which he had long reflected.

Soon after his father's death in 1822, he had thought of collecting William Herschel's various papers, published in the *Philosophical Transactions* over a period of forty years, and bringing them out in a single edition. This would have been a great contribution, because the elder Herschel's publications on double stars and nebulae and his researches on the structure of the Milky Way had been epoch-making works, but several years' solid work would have been required to arrange the papers in a logical sequence, to bring them up to date, and generally to make them into a useful working reference for astronomical research.* The enterprise would have meant almost a complete sacrifice of John Herschel's own original research for the time being.

After careful consideration, Herschel decided that the best way for him to preserve this intellectual inheritance would be to carry on his father's work. He had already started doing this in his revision of the catalogues of double stars and nebulae, but that had essentially been only repetition, extension, and correction; the work had not been completed in the true sense of the word. William Herschel had surveyed the northern celestial hemisphere and a

---

* The compilation was finally made, with the Royal Society and the Royal Astronomical Society collaborating in its production; *The Scientific Papers of Sir William Herschel*, in two volumes, compiled by J. L. E. Dreyer, was published in London in 1912.

small part of the southern, leaving a serious gap in the inventory of heavenly objects and in the statistical analysis of the distribution of the stars. A large part of the southern sky was virtually unexplored. John Herschel therefore determined to sweep the southern celestial hemisphere with the same instruments and the same methods that his father had used for the northern, and thus to ensure the permanent association of the name Herschel with a complete survey of the heavens.

The plan must have appealed almost equally to his scientific earnestness, his capacity for quick decision, and his great love of travel. The first destination considered for the expedition was Paramatta in Australia, but the observatory there was not equipped with the best instruments, the climate left much to be desired, and the length of the journey was too great. Herschel finally decided on Cape Town, South Africa, which seemed promising in every respect. Margaret's brother Dr. Duncan Stewart had gone there from India. Letters were exchanged with him to settle the important points, and soon all arrangements were made.

In the meantime, Herschel's plan had been discussed by his friends and in the Royal Society. The Duke of Sussex, then president, offered funds from the Society to finance the entire expedition, which he considered an enterprise of national importance. The British Admiralty also offered free passage to Cape Town. Herschel refused both offers. He regarded his journey as a purely private undertaking and wanted no public support for it. Once again, as in his previous scientific activities, he wished to be entirely independent. He did not wish to bask in the glory of being a national scientific figure, but only to operate unnoticed and work out the results of his researches in peace. Fortunately his economic situation enabled him to take this lofty attitude.

Nothing now stood in the way of the journey, but Herschel delayed it out of consideration for his aged mother, who would probably not have lived to see her son again after an absence of several years. The preparations were therefore broken off, other enterprises and commitments took up his time, and the journey to the Cape receded into the background. In 1831 he received a high honor: the King dubbed him a Knight of the Royal Hanoverian Guelphic Order, an award which his father had not received until his old age.

Some important family events occurred in the years leading up to the great expedition. On March 31, 1830, a daughter was born to the Herschels and christened Caroline Emilia Mary, the Caroline in honor of Herschel's famous aunt. Now eighty years old, the first Caroline Herschel took a lively share in the joy occasioned by the event, writing delightedly to her nephew: "So I am to be godmother! With all my heart! I am now so enured to receiving honours in my old age, that I take them all upon me without blushing." [40]

In 1831 another daughter, Isabella, was born, and in 1833 a son, William James. Between these two happy events a sad one occurred: Mary Herschel died on January 4, 1832, at the age of eighty-three, in her house at Slough. She was laid to rest next to her husband in the little village church at Upton. The only surviving member of the older generation was Caroline Herschel, still hale and cheerful in her small apartment at Hanover, but now living more in the past than in the present.

Soon after Mary Herschel's death, her son gave up his London house and moved with his family to Slough, where some alterations had to be made in the old house. The studies and libraries were converted into bright living rooms and sunny nurseries, and the silent gray old house and large walled garden were soon lively with the sounds of children.

The 40-foot telescope, whose mounting on the large

lawn overshadowed the roof of the house, was abandoned. Trained by his father in the art of grinding and polishing mirrors, Herschel was working on a new primary mirror for the 20-foot reflector, which he intended to take with him as his main instrument on his journey to the Cape. Preparations for this great enterprise were again pursued with vigor. The necessary instruments were thoroughly overhauled and the working program laid down in detail. In addition, Herschel was continuing his current astronomical observations and completing the catalogues of double stars and nebulae. Presently it became evident that he would not be able to start on the journey until the following year. In May 1832, he wrote to his friend Captain William Henry Smyth: "I find I must give up all thoughts of starting for the Cape this year. Well!—I don't regret it. Anything is better than hurry and it would be provoking indeed to arrive there and find anything essential omitted or insufficiently done." [41]

In June 1832, he paid a farewell visit to Caroline Herschel at Hanover. "I found my aunt wonderfully well," he wrote to his wife, "and very nicely and comfortably lodged, and we have since been on the full trot. She runs about the town with me and skips up her two flights of stairs as wonderfully fresh at least as *some folks* I could name who are not a fourth of her age. . . . In the morning till eleven or twelve she is dull and weary, but as the day advances she gains life, and is quite 'fresh and funny' at ten or eleven p.m., and sings old rhymes, nay, even dances! to the great delight of all who see her. . . . It was only this evening that, escaping from a party at Mrs. Beckedorff's, I was able to indulge in what my soul has been yearning for ever since I came here—a solitary ramble out of town, among the meadows which border the Leine-strom, from which the old, sombre-looking Markt-thurm and the three beautiful lanthorn-steeples of Hanover are seen as in the little picture I have often looked at with a sort of mysterious wonder when a boy as that strange place in foreign parts that my father and uncle used to talk so much about, and so familiarly. The *likeness* is correct, and I soon found the point of view." [42]

At last the time came to embark on the great journey. The house at Slough was again abandoned, this time to be empty for years. Trunks and suitcases were packed, and the instruments, safely encased in large boxes, were taken by boat down the Thames from Windsor to London, where they were stowed in the hold of the *Mountstuart Elphinstone*, a vessel of 611 tons. The last farewell visits were paid and on November 10, 1833, the Herschel family, whose youngest member, William James, was just six months old, set out for Portsmouth to embark on the ship from London, just arrived by way of the Channel.

Up to the last moment, Herschel had been burning the midnight oil to complete whatever scientific work he could before his departure. "The last proof sheets of my nebulae paper," he writes to his aunt from Portsmouth, "left my hands the night I left London, and yesterday I got twelve copies to take to the Cape. . . . My observations on the satellites of Uranus, which confirm my father's results, are sent to be put in course of publication last night." [43]

On November 13 they finally set sail on the long voyage to Cape Town, which was to last nine weeks and two days and to usher in one of the happiest and most untroubled periods of Herschel's life.

Observatory House, Slough, where John Herschel was born

Sir William Herschel (portrait by Lemuel Francis Abbott, 1785)

'Lady Herschel
(from a miniature painted on ivory by J. Kernan, 1805)

John Herschel at the age of 39 (drawing by H. W. Pickersgill)

Nov. 9 1816

γ (Fl. 5) Andromeda. Position $9R + 25P + 0.3Z =$
$= 25° 11'.55$ nf not accurate

2d Measure $8R + 85.5P + 0.3Z = 19° 58'.80$ nf
more accurate. Clouds prevented a
repetition.

Nov. 19. 1816

Zero of the Cross wire Micrometer. By a mean of ten
measures taken with one side of the moveable hair (by
candle-light) zero = 0.00 ___ a mean of ten on the
other side gave zero = 0.610.
Mean zero = 0.305 to be added } shuts at
central ob   5

Nov. 20. 1816

Zero of Field's small parallel wire Micr. By a mean of
ten obs. It shuts at 18.825 ∴ Zero = − 18.825
Zero of Nairne's large do do. By mean of 10 obs.
It shuts at 0.950 ∴ zero = − 0.950. N.B. This
has an adjustable index
Both these zeros are central where the wire just
covers the other.

Nov. 23.

7 feet Reflector. Power 270 Cross wire Micron. of 100 feet
β Cygni (Fl.) Double at a considerable distance, unequal large white
a yellowish wh. Small blue
Position $9R + 32.5P + 0.3Z = 29° 52'.8$ nf
2d Meas. $9 Rev 39P + 0.3Z = 31° 10'$ nf
3d $9 Rev 32.5P + 0.3Z = 29° 52'.8$ nf

Orion (1 Fl) coarse double star, three distance of β Cygni uneq
both white

Nebula in Andromeda 31 Connois des Temps. Extended, gradually
much brighter in the middle, perfectly nebulous in its appearance
& no sign of Stars. — Visible to naked eye (N.B. Double Eyeglass)

32 Connois double Eyeglass, much brighter in middle, in which
there is the semblance of a Star. — Power 118 fancied
resolvable, but faint — Power 115 very faint no appearance of Stars

36 Connois a pretty rich cluster of moderately large Stars, figure
irregular, the principal Stars form a kind of

Facsimile of a page from John Herschel's first astronomical notebook

Margaret Herschel (miniature by A. E. Chalon, R.A., 1830)

Caroline Lucretia Herschel
(portrait by Melchior Gommar Tielemann, 1829)

Feldhausen, John Herschel's home near Cape Town, South Africa
(photograph taken by Caroline Herschel Gordon in 1898)

The 20-foot reflector at Feldhausen
(from a camera-lucida sketch by John Herschel—lithograph G. H. Ford)

The 40-foot telescope at Slough
(from the first negative taken
on glass by John Herschel,
September, 1839)

Dismantling the 40-foot telescope
(camera-lucida drawing by John Herschel, 1840)

Collingwood, the Herschel home at Hawkhurst, Kent

Sir John Herschel, about the time he became Master of the Mint
(from a photograph by J. Dudgen, 1853)

Portrait of Isabella, John
Herschel's second daughter,
taken by Louisa, his third
daughter, in 1857

The Herschel sons, 1883;
left to right: John,
Alexander Stewart, William James

The obelisk erected by friends of John Herschel at the site
of the observatory at Feldhausen

4

At the Cape of Good Hope

✦

✦

✦

"1st Jan., 1834. Commenced the New Year in lat. 29° south, long. 11° west on board of the *Mt. Stewart Elphinstone* [*sic*], expecting to arrive at the Cape of Good Hope in ten days or a fortnight (which God grant!). Hitherto all has gone on most smoothly—no adventures—no accidents—no foul weather—no Bay of Biscay—no long fretting calms at the equator—but handed over regularly from one fair wind to another with little more interval than enough to make us acknowledge the powers under which Neptune's Empire stands."[1]

Thus Herschel's diary ushers in the year 1834 and the expedition that is in many ways unparalleled in the history of astronomy, in regard to both the methods by which it was carried out and the number and importance of the results it produced.

When John Herschel decided to extend to the southern sky his father's great work of surveying the northern celestial hemisphere, he faced a tract of virgin scientific territory comparable to that which William Herschel had attacked fifty years earlier. The English astronomer Edmond Halley (1656–1742), who discovered the comet that bears his name, had observed the southern sky from St. Helena between 1676 and 1678, and contributions to cataloguing it had been made by the French cleric and astronomer Nicolas Louis de Lacaille (1713–1762) in his posthumous *Caelum Stelliferum Australe* and in the lists of stars and nebulae by the Scottish soldier and astronomer Thomas Makdougall Brisbane (1773–1860) and by the British astronomer James Dunlop (1795–1848), who worked at Paramatta. These, however, had been only humble beginnings, undertaken sporadically, often with inadequate means, and containing appreciable errors. For example, of the 600 nebulous objects listed by Dunlop in his catalogue published in *Philosophical Transactions* in 1826, Herschel was able to find only about a third, because the positions were so unreliable and the descriptions so inaccurate. At the time, there were no large, well-equipped observatories in the southern hemisphere. The only ones of any importance at all were at Paramatta, Australia, and Cape Town, South Africa, and these had only small instruments, used mainly for meridian observations, and in no way comparable in performance to the 20-foot telescope which Herschel set up at the Cape.

Thus when Herschel arrived at Cape Town on January 16, 1834, he faced a formidable undertaking. After the apparatus and baggage had been unloaded, which took several days, he began to look for a suitable place to live with his family for the several years of his stay at the Cape. He found and rented a spacious property called Feldhausen, a Dutch farmhouse about six miles southeast of Cape Town, picturesquely situated at the foot of Table Mountain and surrounded by large orchards and a park-like grove of oak and fir trees. The locality had a very good climate, and observing conditions were much better at the Cape than in England. Herschel found that the clearest and most transparent nights were those of the cool season (from May to October), especially just after the heavy rainfalls that are common there at this time of year. "On these occasions the tranquillity of the images and sharpness of vision is such, that hardly any limit is set to magnifying power but what the aberrations of the specula necessitates." [2]

Herschel immediately set about the erection of an observatory at Feldhausen. Bringing the heavy chests containing the various parts of the instruments out from Cape Town was a major operation, as was putting up the telescope with its enormous wooden mounting. Herschel had brought with him from England the mechanic John Stone, who had assisted him during his survey of the northern sky at Slough. Stone supervised the work of the native laborers and afterward provided astronomical and engineering assistance in the use and maintenance of the telescopes.

The erection of the observatory proceeded rapidly. "On the 22nd of February," reports Herschel in the preface to his report of the expedition, familiarly known as *Results*, "I was enabled to gratify my curiosity by a view of a Crucis, the nebula about η Argus, and some other remarkable objects, in the 20 ft. reflector, and, on the night of the 5th of March, to commence a regular course of sweeping." [3]

Like most of the Herschel telescopes, the 20-foot reflector was constructed on the Newtonian principle as described in Chapter 1. It was suspended by a system of ropes from a framework mounted on movable rollers. A movable platform gave the observer access to the eyepiece in any position. Of the three interchangeable mirrors constructed for this telescope, one had been made by William Herschel, another had been ground and polished by father and son together, and the third had been produced by John Herschel on his own. All three mirrors had the same aperture of 18¼ inches and the same focal length of 20 feet, so that they were identical in their optical performance.

Having more than one mirror available proved very helpful. Herschel did not need to interrupt his observations when the mirror in current use became tarnished and needed repolishing, which happened frequently. Often a mirror deteriorated so much in as little as a week's time that it was useless for observation and Herschel had to work for several hours at the polishing machine to restore its reflectivity. The rapid corrosion of the mirror surface was probably caused by sea-salt nuclei blown in by the winds.

Apart from the large reflector, Herschel had also brought out a second telescope, an equatorially mounted refractor of 7-foot focal length. This was the instrument he and South had used for double-star measurements. Whereas the reflector was mounted in the open air, the sensitive graduated circles and micrometers of the refractor had to be shielded from the weather, and it was accordingly housed in a separate building with a sliding roof. Adjusting this telescope took several months. Herschel started using it for systematic observations on May 2, shortly after the family had finally moved into Feldhausen. He began with the fine double star Alpha Centauri, one of the brightest stars in the sky. The refractor was chiefly used to make measurements of double stars and other celestial objects and to check and supplement the results obtained with the reflector by providing exact positions of the relevant objects on the celestial sphere.

The reflector was used for a systematic survey of the heavens in the manner of William Herschel's sweeps of the northern hemisphere. Starting at the celestial equator, Herschel searched for new objects in a series of zones 3° wide in declination. Every star cluster, nebula, and binary was carefully observed, described and measured, and its coordinates in right ascension and polar distance were determined with the aid of the equatorial refractor.

The results of this work, among others, were recorded in two catalogues: a catalogue of nebulae containing 1707 objects (of which only 439 were known previously) and a list of double stars containing 2102 binary pairs. The latter number is particularly impressive in view of Herschel's finding that the southern hemisphere is appreciably less well endowed than the northern with bright double stars.

Simultaneously with these surveys, he undertook counts of the number of stars visible in the field of view of the telescope in no less than 3000 sections of the sky. The total was 68,948. These "star gauges," previously undertaken by William Herschel, were intended to provide information on the distribution of the stars of the Milky Way in space. In the course of these investigations, John Herschel was led to form a picture of the structure of the galaxy that differed somewhat from his father's views. While the latter had considered it to be a flat disk or lens-shaped system, John believed it to be a ringlike structure which included the solar system and all the visible stars. According to his conception, the sun with its surrounding planets was not exactly at the center or hub of the wheel, but closer to the southern parts of the Milky Way. The latter appear to be brighter and more richly populated than the northern parts. From the fact that the Milky Way does not exactly describe a great circle on the celestial sphere, Herschel deduced that the solar system must be situated at some distance away from the plane of symmetry of the galaxy. He seems to have been firmly convinced that its structure was annular. In 1836 he wrote to William Rowan Hamilton: "It is impossible to resist the conviction that the Milky Way is not a stratum but a ring." [4]

He was immediately struck by the fact that the stars in his galactic ring were by no means uniformly distributed. In some places—for example in the region of the Coalsack, which is a practically starless area in the middle of the Milky Way—there were striking gaps in the luminous belt of stars, while in others the stars were so numerous that they merged together in bright clouds and stood out in relief against the dark background of the sky. Not until much later did astronomers realize that these starless areas in the Milky Way are not real holes, but an apparent effect brought about by opaque clouds of dust and gas.

Herschel estimated that altogether about 5.3 million stars in both celestial hemispheres could be seen with a 20-foot telescope of 18 inches aperture used with a magnifying power of 180. However, he wrote in *Results*: "That the actual number is much greater there can be little doubt, when we consider that large tracts of the Milky Way exist so crowded as to defy counting the gauges, not by reason of the smallness of the stars, but their number." [5]

In his "gauges," Herschel came to essentially the same conclusion that his father had reached with regard to the northern sky: "Nothing can be more striking than the gradual but rapid increase of density on either side of the Milky Way as we approach its course . . ." [6] The gauge fields were chosen so as to be nearly uniformly distributed over the sky. They were 10 minutes (= 2° 30′ of arc on the equator) apart in right ascension and 1° 30′ in polar distance.

An object of lengthy research was the determination of magnitudes—that is, the relative apparent brightnesses of the stars. William Herschel had given estimates of the magnitudes of about 2000 northern stars in four catalogues.[7] In each case he had also listed two or three nearby stars having the same apparent brightness, or nearly so, so as to provide later observers with standards on the basis of which any variations or fluctuations in brightness could be detected; this was intended to facilitate the discovery of variable stars, of which only a few were known at the time. This procedure was extended and modified by John Herschel, who developed what he called the "method of sequences." To make the estimations of brightness as accurately as possible, he invented a simple and ingenious device which can be claimed to be the first stellar photometer. On March 5, 1836, he wrote in his diary: "Tried a project for comparing stars with moon by total reflection at the base of a prism." [8]

His instrument, which he called an "astrometer," was constructed as follows: A totally reflecting prism was mounted on a wooden platform. By suitable rotation of the prism, the light of the moon could be directed to the eye in such a way as to be parallel to (that is, to appear to come from the same direction as) the light from the star whose brightness was to be measured. A lens of short focal length, adjustably mounted between the prism and the eye, reduced the image of the moon to the size of a point, so that it could be used as an artificial star, so to speak. The brightness of the artificial star could be varied by altering the distance between the lens and the eye until the real and the artificial star appeared to be equally bright. In this way the distance from the eye to the lens gave a measure of the apparent magnitude for any star.

In this way, Herschel determined the relative brightnesses of 191 of the brightest stars in both hemispheres. Alpha Centauri, which he used as a reference star and standard, he found to be about 27,000 times fainter than the full moon. Since Wollaston had determined the ratio of the brightness of the full moon to that of the sun to be 1:800,000, Herschel was now in a position to compare the stellar brightnesses determined by him with that of the sun. For those stars whose distances were known in Herschel's time (the first stellar parallax was not determined until 1838), it was then possible to determine the intrinsic luminosity reduced to the distance of the sun. Alpha Centauri proved to be four times as luminous as the sun, Vega (Alpha Lyrae) forty times, and Arcturus (Alpha Boötis) two hundred times.

Herschel's "lunar method" was subject to certain sources of error. In particular, scattered moonlight caused a brightening of the sky background which had a disturbing effect on the exact determination of stellar brightnesses. Herschel does mention the idea of using the planet Jupiter instead of the moon, as a comparison source, but he does not seem to have tried this scheme, or the very similar idea of using an artificial light source as the German astrophysicist Johann Karl Friedrich Zöllner (1834–1882) did afterward in his photometer.

However, Herschel's photometric investigations were not based on his astrometer measurements, but on his step-method, which was carried out as follows. From the star charts made by the German astronomer Johann Elert Bode (1747–1826), triangular fields were selected and in these all stars visible to the naked eye were arranged in sequences. This was done by writing down a list of stars having very marked differences in brightness between them, leaving enough space between successive entries so that more stars could be interpolated. In this way Herschel obtained a sequence in which the finest detectable gradations in brightness were shown. A sequence was complete when all the gaps were filled. Altogether he carried out about 2300 estimations of brightness in 46 sequences covering some 500 stars.

Star clusters and nebulae in the southern sky occupied a dominating position in Herschel's research program. His catalogue of more than 1700 nebulae, arranged in the form of a table, occupies 80 pages in *Results*. As in his earlier catalogues, the nebulae are arranged in order of right ascension. Polar distances are given, rather than declinations, which are now customary. There are short notes on individual objects to help in identification.

Herschel also carried out careful investigations of the distribution of nebulae over the celestial sphere. By a statistical method, consisting of counts of objects at various ascensions, he found a striking decrease in average numbers in the neighborhood of the Milky Way, whereas in

regions remote from the Milky Way—the galactic poles—
a marked concentration of nebulae was observed. In particular, the constellation of Virgo, situated near the North
Galactic Pole, contained an extraordinarily rich cluster of
nebulae, but the neighborhood of the South Pole also had
a large number, for example, in the Pisces region. This
discovery was of great cosmological significance, because
it seemed to show that the realm of the nebulae and the
Milky Way were two distinct systems. However, the
brighter star clouds and open star clusters almost always
occurred in the immediate neighborhood of the Milky
Way and on either side of the bright band of stars and
were virtually absent at the galactic poles.

In the state of cosmology at the time, Herschel was
unable to explain the accumulation of nebulae near the
galactic poles. In fact, the explanation was not discovered
until a hundred years later. It is now known that the
Milky Way system is an enormous spiral galaxy of stars,
whose central plane contains vast masses of cosmic dust
that obscure our outward view. Since many of the objects
described by Herschel as "nebulae" are in fact galaxies
similar to our own Milky Way system, they can only be
detected in those directions in which the dark cosmic
clouds of our own galaxy do not block the view. This is
notably the case in the direction of the galactic poles.
Thus the concentration of extragalactic nebulae in these
regions is only apparent.

Certain particular objects were given an unusual degree
of attention in Herschel's observations. Among these the
most important are the Magellanic Clouds, also known as
the Cape Clouds and the Nubeculae. These are two large
star clouds with surface areas of 40 and 10 square degrees
respectively. In brightness they match the brighter portions of the Milky Way, but both are completely detached
from the Milky Way belt. The Small Magellanic Cloud is
situated in a virtually starless region in the constellation

of Tucana. On its edge, but separated from it, is the globular star cluster 47 Tucanae, a striking object. Its densest central parts merge into a nebulous haze which Herschel was unable to resolve into stars. He discovered a wealth of star clusters and nebulae of all kinds in this part of the sky.

Still more impressive and interesting was the Large Magellanic Cloud, Nubecula Major. "In the number and variety of . . . objects, and in general complexity of structure, it far surpasses the Lesser Nubecula . . ." [9] Although Herschel counted only 40 star clusters and nebulae in the latter, the Large Cloud contained no less than 278 objects, in addition to about 50 or 60 "outriders." "It is evident," he wrote, "that the nubeculae are to be regarded as systems sui generis, and which have no analogues in our hemisphere." [10] He drew up a catalogue of the various stars, nebulae, and clusters observed in the two Magellanic Clouds, containing altogether 919 objects in the Large Cloud and 244 in the Small, arranged in order of right ascension and with data on brightnesses. This information was intended to form the basis of a precise map of the two Clouds. He produced such a map for the Large Cloud, a masterpiece of celestial topography, although he modestly entitled it "First approximation to a Chart of the Nubecula Major or Greater Magellanic Cloud." It is much more than a "first approximation"; in fact it is probably the best hand-drawn map of the Large Magellanic Cloud ever to have been prepared by an observer directly from his results at the telescope.

Next to the two Magellanic Clouds, Herschel was most fascinated by the extensive nebulous region surrounding the star Eta Argus.✦ The nebula is surrounded by cosmic

---

✦ The large constellation Argo (the Celestial Ship), which lies near the Southern Milky Way and abuts on Crux (the Southern Cross), has now, partly on Herschel's initiative (see Chapter 5), been divided into four parts: Carina (the keel), Vela (the sails), Puppis (the poop), Pyx (the compass). The object which Herschel called Eta Argus, or sometimes Eta Roboris, is now called Eta Carinae.

dust clouds and thus stands out especially clearly from the other parts of the Milky Way as a result of the contrast. According to Herschel's observations, the nebulous region covers a larger area than the Large Magellanic Cloud—47 square degrees. Estimates based on star gauges gave a total of 147,500 stars bright enough to be observable with the 20-foot telescope. Thus an extraordinarily rich Milky Way field presented itself.

In the midst of this star cloud is situated Eta Argus or Eta Carinae, a fascinating object, as Herschel found in the course of his observations. Its apparent magnitude had been given as 4 in 1677 by Halley, the first systematic observer of the southern sky, whereas Brisbane, the Reverend Fearon Fallows (1789–1831), first director of the Cape Observatory, and others had classed it as a second-magnitude star in their catalogues. Herschel himself, who measured the brightnesses of stars with especial care, found it to be between the first and second magnitude. So it remained until the end of 1837. Then within a fortnight the star suddenly acquired a brilliance approaching that of Alpha Centauri and noticeably exceeding that of the bright first-magnitude star Rigel. By mid-January 1838, however, its brightness had already diminished markedly. "On the 20th [of January]," he writes, "it was visibly diminished—now much less than *a* Centauri and not *much* greater than Rigel. The change is palpable." [11] Herschel was unable to follow the subsequent course of the light curve, because he embarked on his journey home to England in March 1838. At the time of his last observation, on April 14, presumably made on shipboard, Eta Carinae had reached about the brightness of the star Aldebaran in Taurus. Herschel's results were confirmed by other observers, among them Her Majesty's Astronomer and Director of the Cape Observatory, Thomas Maclear (1794–1879). All were equally struck by the very sudden rise of the light curve to a maximum and its equally speedy decline. Maclear even compared the maximum brightness to that of Sirius and found it to be only slightly less.

Herschel worked for months preparing a large special chart of this region around Eta Carinae, based on a list of 1216 stars in the area down to seventeenth magnitude. Using Eta Carinae as the origin of a coordinate system, their positions were measured to an accuracy of 1/10 second of arc, an arduous task in which South's 7-foot equatorial gave good service. Herschel writes of this lengthy and tedious piece of work: "Frequently, while working at the telescope on these skeletons, a sensation of despair would arise of ever being able to transfer to paper, with even tolerable correctness, their endless details." [12] The Carina nebula derives its characteristic appearance from a striking dark cloud that occurs in its most richly populated part, immediately adjacent to Eta Carinae. In its outlines this dark nebula resembles a key-hole, and hence the whole region is sometimes known as the Keyhole Nebula.

Herschel also found peculiar shapes to be common among other nebulae. He preserved the most striking shapes in small sketch maps, for example, a map of the Trifid Nebula, a bright gaseous nebula that is split into three clearly separated parts by interposed tongues of dark matter. Many other nebulae appeared as more or less flattened ellipses with central condensations, going from an almost lenticular shape to that of a perfect circle. It is now known that most of these nebulae are remote galaxies, which appear in a great variety of perspectives according to the angle at which they are viewed. Seen "edge on," they resemble flattened nebulous spindles, which open out into ellipses and finally into circles as the angle between the line of sight and the equatorial plane of the nebula increases. Herschel did not have a clear idea of the spiral character of many of the galaxies. This discovery was reserved for his fellow countryman William Parsons, third

Earl of Rosse (1800–1867), whose telescope Leviathan exceeded in performance even the Herschels' giant instruments. In two cases, however, Herschel speaks of "falcated" (sickle-like) nebulae, and this certainly refers to the spiral arms that emerge from their nuclei. His sketch of the Large Magellanic Cloud also gives a slight indication of spiral structure. That Cloud is, in fact, a galaxy close by our own Milky Way, from which it is separated by about 170,000 light-years, and together with the Small Cloud and the Milky Way it forms part of a local group of galaxies.

Herschel was also fascinated by the globular star cluster 47 Tucanae, an object of unusual beauty and regularity, which can be seen with the naked eye as a fuzzy spot of light in the immediate vicinity of the Small Magellanic Cloud. Humboldt, who had observed it repeatedly during his expedition to the Andes, had at first taken it to be a comet because of its remarkable size and brightness. Herschel was struck by the perfect geometrical regularity of the figure of the cluster. He thought he could distinguish in it three distinct zones of star density—the densest region, the nucleus, shone with a rosy glow, whereas the middle and outer zones had a whitish coloration. This effect, however, does not seem to have been detected by other observers.

One of the most impressive pieces of celestial cartography ever produced is Herschel's detailed drawing of the Great Nebula in Orion. By far the most striking of all nebulae, this can be observed from both northern and southern latitudes, as it lies only a few degrees south of the celestial equator. Herschel's South African view was much better than that obtainable in England, for at Feldhausen the Orion nebula culminated at an altitude of 60° above the northern horizon, much farther from the mists which dim objects at low altitudes.

For his 1826 monograph on the Orion nebula, Herschel

had attempted to record its details in a drawing. Viewing the nebula in the clear sky of Feldhausen, however, he realized that there was very much more to the nebula than he had been able to detect through the less favorable atmosphere of England. "Yet the first glance . . . sufficed to convince me of the necessity of executing a redelineation of it." [13]

On many observing nights, spread over several years, Herschel produced a number of outline sketches of different parts of the nebula. Each sketch was carefully checked and compared with positions measured at the telescope. A list of 150 stars in the field of the nebula was used to provide reference points for the production of the final chart, which was complicated by the great range in brightness of the various portions of the nebula.

Herschel's chart of the Orion nebula is not the only one of its kind. Since his first drawing of the nebula in 1826, there had been four further attempts to record the confusing wealth of detail. These representations show considerable discrepancies in some features, a fact which Herschel did not ascribe to real changes in the structure of the nebula, but to the difficulty of giving a completely objective picture of such a complex object. "Now there is only one such particular," he writes, "on which I am at all inclined to insist as evidence of change; viz. in respect of the situation and form of the 'nebula oblongata,' which my figure of 1824 represents as a tolerably regular oval extended very nearly in a right line, or at most but a very little curved upwards between the two stars Chi and Kappa of the catalogue. Comparing this with its present appearance . . . it seems hardly possible to avoid the conclusion of some sensible alteration having taken place." [14]

Although the observation of nebulae and star clusters occupied a considerable portion of Herschel's astronomical researches, his writings do not contain any general theo-

retical view or interpretation of these objects. He seems to
have been at least somewhat critical of his father's hypothesis that there were vast numbers of cosmic gaseous
masses that produced stars by a gradual process of con-
densation. John Herschel assumed that many nebulae cannot be resolved into stars merely because of the high
degree of overcrowding in their central parts, or, in other
words, that the majority of "nebulae" are stellar systems
that cannot be resolved owing to their high density. In
Outlines of Astronomy, he raised the question "whether
there be really any essential physical distinction between
nebulae and clusters of stars, at least in the nature of the
matter of which they consist." [15] The distinction among
nebulae which are easy, difficult, or impossible to resolve
into stars he regarded as a purely accidental one, resulting
from the combination of the optical performance of the
telescope with the number, size, and overcrowding of the
stars in a nebula. Thus a small instrument can resolve
only the brightest nebulae into stars, whereas in a large
telescope many objects that with less powerful optical
means seemed like unresolved nebulae immediately stand
out as clusters of stars densely crowded together. This
assumption received strong support from the results secured with Lord Rosse's reflector (diameter 72 inches),
completed in 1845. With this giant instrument there
seemed to be no limit to the resolution of nebulae into
stars. Even in the Orion nebula, Lord Rosse thought he
saw indications of starlike structure. Not until the spectroscope was applied to the study of this and other nebulae
was it realized that the sky contains genuine nebulae—
that is, clouds of cosmic gas. This discovery confirmed
what William Herschel had suspected in 1791 when he
spoke of "a shining fluid, of a nature totally unknown to
us" and advanced the view that the wealth of nebulous
material in the heavens was so great that it "must exceed
all imagination." [16]

The main purposes of John Herschel's expedition to the Cape of Good Hope were first to extend the star gauges carried out by William Herschel in the northern hemisphere, and, secondarily, to carry out, again supplementing and computing William Herschel's work, a systematic survey of the southern hemisphere for double stars, nebulae, and other celestial objects of particular interest. The star gauges showed the same result as in the northern hemisphere—the number of stars in single gauge fields increased rapidly and regularly as one approached the Milky Way. William Herschel had deduced from this that the galaxy had a disklike shape with an axial ratio of about 1:5, whereas John inclined to the hypothesis of an annular structure. The sky surveys revealed more than 1700 nebulae and clusters and 2100 double stars. The reductions of the observations recorded in the two catalogues carried out by John Herschel involved a great deal of monotonous arithmetic, and kept him occupied for several years after his return from the Cape. A further result of the expedition was a series of researches in photometry, which were the first attempt to carry out such work in the southern sky. His various individual studies of nebulae and star clusters, including his monographs and charts of the two Magellanic Clouds, the nebulous region surrounding Eta Carinae, and the Orion nebula, were important enough in themselves to have justified the expedition to the Cape. It has often been asserted that John Herschel lacked the creative genius and scientific pioneering spirit of his illustrious father, but the wealth of results from the Cape expedition shows the brilliant achievements of which John Herschel was capable. He was not inferior to his father in observational industry and endurance, and in scientific precision and versatility he actually exceeded him by a wide margin.

John Herschel's researches in sidereal astronomy were not the only results of his stay at Feldhausen. During those four years he also did important work in other fields of practical astronomy. At the end of 1835 and the beginning of 1836 he had the good fortune of being able to make detailed observations of the return of Halley's comet. On October 28, 1835, he noted in his diary: "Knocked up a temporary stand for the 7 feet equatorial telescope, dismounted it and carted it out to the first sand hills on the flats; there erected it just at sunset and was rewarded with the first glorious sight of Halley's comet!" [17]

On the following evening Herschel, using the reflector, managed to observe the comet from Feldhausen. In order to be able to observe it, despite its low elevation just above the horizon, he had some oak trees cut down because they blocked the view.

The comet was readily observable in the bright evening twilight until November 10, after which its period of visibility rapidly diminished as it approached the sun. Herschel rediscovered it in the dawn twilight on January 26, 1836, in the constellation of Scorpio. It had in the meantime gone through perihelion and was now· receding, so that it became gradually fainter and more indistinct. The last occasion when it was visible in the reflector was May 5.

Herschel made an important contribution to the records of Halley's comet and to cometary studies in general by describing in detail the comet and its changing forms in the course of the perihelion passage. Some fundamental remarks are directed to the question of what forces operate on the comet's tail at the time of perihelion. The tails of comets consist of extremely rarefied gaseous material which evaporates from the nucleus when the comet approaches the sun. The tail is always directed away from the sun, because of the pressure of the "solar wind." Herschel's ideas are based on the assumption of the presence of positive and negative electric charges. He wrote in the *Results*: "I cannot help remarking that the conception of a high degree of electrical excitement in the matter of the tail . . . would satisfy most of the essential conditions of the problem. That the sun's heat in perihelio does actually vaporize a portion of the cometic matter, there can no longer, I think, be any reasonable doubt. That in such vaporization, a separation of the two electricities should be effected, the nucleus becoming (suppose) negative, and the tail positive is in accordance with many physical facts." [18] Bessel had also expressed the view that the phenomenon of cometary tails could be attributed to electrical effects.

Herschel completed his article on Halley's comet with a number of sketches which give an excellent idea of the structural changes which it underwent at various phases of its passage.

In the period between 1835 and 1837, Herschel made many observations of the ringed planet Saturn and its seven then known satellites, the two innermost of which had been discovered by William Herschel with the 40-foot telescope at Slough. Conditions were especially favorable for observation during those years because the planes of the ring and of the orbits of the two satellites were then inclined at a considerable angle to the line of sight. Furthermore, the two satellites found by William Herschel —Mimas and Enceladus—had not been observed again since 1789. John Herschel himself had searched for them in vain from England and was unable to find them again until he was under the clear skies of South Africa. Afterward, these two satellites, which are the closest to the planet * and very faint, were again lost sight of until they were rediscovered in 1846 by the English astronomer William Lassell (1799–1880). Even now, they are among the most difficult objects to observe in the solar system.

Herschel measured the position angles and apparent magnitudes of all seven satellites of Saturn and presented the results in the form of tables. These provided valuable

---

* Except for the satellite Janus discovered in 1966.

data for improved determination of their orbits, of which, especially the fainter ones, very little was known at the time.

The last chapter of the *Results* contains a contribution of genuine importance to solar research and the study of sunspots. Good opportunities for observing were provided by the years 1836 and 1837 when the sun was particularly active. Whenever time and weather permitted, Herschel carried out counts of sunspot numbers and made drawings of the solar surface by projecting its image onto a paper screen with a telescope; in this way he was able to record the exact positions of the spots on the disk. Remarkable single spots or groups he drew in detail from observations with an eye-piece of magnifying power between 105 and 179. Some of these drawings are given in an appendix to the *Results*. Anyone who has attempted to draw sunspots at the telescope will realize what a hard task this is, because sunspots have extraordinarily complicated shapes and their outlines are hard to define. The light-gray outer zone, known as the penumbra, which surrounds the dark spot proper and often has a peculiar jagged and twisted appearance like a whirlpool, is especially difficult to represent faithfully.

Herschel did not confine himself to description of sunspots; he also attempted to develop a theory of the physical causes of the phenomena that occur on the solar disk. Here he shows some of the creative powers of imagination so prominent in his father's work, but in many of the son's papers too often concealed behind dry scientific arguments couched in purely technical terms. In his theory of the sun, John Herschel begins by assuming that the appearance of sunspots must be closely related to solar rotation, because the spots are mostly confined to certain distinct zones of heliographic latitude. Spots appear on either side of the solar equator out to latitudes of 35° north or south but are not observed either at the equator itself or at higher latitudes including the poles. Herschel concludes from this that, as a result of an uneven distribution of heat in the solar atmosphere, vast streaming motions take place, leading to an exchange of heat between the poles and the equator, and that the photospheric gases which surround the body of the sun proper are driven by its rotation into a streaming motion which he compares to the trade winds on either side of the equator of the earth. In these confluent masses large vortices form, resembling tornadoes and hurricanes on the earth, but on a vastly greater scale. Expressed in simple terms, "holes" appear in the solar atmosphere and through these we can see the core of the sun itself. According to Herschel, this core was a dark, solid body and the source of all its heat energy was found in its atmosphere. John Herschel took over the hypothesis of a nonluminous solid solar core from his father, who in turn took it from the Scottish astronomer Alexander Wilson (1714–1786). Supported by the authority of the Herschel name, this idea survived well into the nineteenth century and was not finally abandoned until precise physical and chemical analyses of the sun had been made possible by the spectroscope.

Herschel assumed that the solar atmosphere, or rather the outer layer of the sun which produced its heat and light, was extremely distended. As evidence for this he adduced the appreciable diminution in the brightness of the solar disk that takes place as one goes from the center to the edge ("limb-darkening"), an effect that is immediately obvious if one looks at the sun through a piece of smoked glass or projects it onto a sheet of white paper. He regarded the appearance of three large prominences during the total solar eclipse of July 8, 1842 as further evidence for his hypothesis, because he considered them to be enormous cloud formations "which must have floated in and been sustained by an exterior transparent atmosphere." [19] In view of the rotation of the sun about its axis, Herschel reasoned, the solar atmosphere must have the shape of an ellipsoid of revolution. From this he concluded further that the output of heat into space from the polar regions of the sun, where the atmosphere was relatively thin, must be greater than that from the equatorial zone, where it was more extended. This result finally led him to assume a temperature difference which would account for the thermal streaming effects that gave rise to sunspots.

This theory of Herschel's is no longer considered valid; although the outer layer of the sun does indeed take up the form of an ellipsoid as a result of its rotation, the departure from an exact sphere is extremely slight and not nearly enough to produce an appreciable temperature gradient between the equator and the poles. However, the supposition that there is a close connection between the sun's rotation and the fact that sunspots occur in particular zones of latitude is perfectly reasonable.

An interesting suggestion, hidden in a footnote in Herschel's discussion of sunspots,[20] was a proposal for a permanent watch over sunspot activity, to be carried out from various observatories, so as to make available a continuous description of events on the solar surface. He also drew attention to the possibility of using photography for the purpose, thus making the photographic plate into an objective chronicler of the sun.

This proposal at first received little attention, but in 1858 the English astronomer Warren de la Rue (1815–1889) began to photograph the sun systematically from his private observatory at Kew, and eventually various major observatories around the globe took up regular solar patrols as part of their working program. Nowadays a considerable number of observatories are devoted entirely to solar research.

Herschel's solar researches also extended to the purely physical field of radiation measurement. During his travels on the European continent, he had measured the heating effect of solar rays with his newly invented actinometer (see Chapter 2). The instrument that he used at the Cape consisted of a cylindrical vessel of sheet-zinc 3.75 inches in diameter and 2.4 inches deep, filled with dark-colored water so as to achieve as complete a degree of absorption as possible. The vessel was placed in a larger metal cylinder and was additionally shielded from indirect radiation by a number of screens. The instrument was exposed to direct solar radiation for a few minutes while the sun was near the zenith, and the temperature of the liquid was then measured with a sensitive thermometer. The measurements were repeated several times and averaged to obtain a measure of the flux of solar radiation incident on a given area in a given interval of time.

The unit of measurement for solar radiation, or "actine," was defined by Herschel "to be that intensity of solar radiation which, if entirely intromitted and absorbed, would suffice to melt one-millionth of a metre in thickness per minute off the surface of a sheet of ice co-extensive in area with the sunbeam, and perpendicularly exposed to it." [21] Herschel estimated that one-third of the incident heat rays is absorbed by the earth's atmosphere, but this estimate is much too low. The total heat radiation of the sun would suffice, Herschel wrote, "to melt a cylinder of ice 184 feet in diameter, and in length extending from that luminary to α Centauri," in one minute.[22]

Although Herschel's actinometric experiments, in which he used a great variety of instruments, have now been largely forgotten, they contributed considerably to the development of systematic research on solar radiation and thus to the foundation of an important branch of astrophysics.

So far, this account of the results of Herschel's researches at the Cape of Good Hope has covered only the astronomical aspect of the expedition. This was by no means the only one, although his *Results* might give that impression.

Herschel had refused the offers of financial support from the British Admiralty and the Royal Society in order to preserve the private character of the enterprise and avoid being committed to a prescribed program. Consequently he was free to decide exactly what, and how much, to do and to make plans in accordance with his own scientific conceptions.

He had the good fortune to find in the Cape Observatory's director, Thomas Maclear, both an able colleague and a warm-hearted, congenial friend. Maclear had arrived at the Cape only ten days before Herschel did, and an active and mutually profitable collaboration soon developed. Maclear provided Herschel with the exact positions of a number of fundamental stars * secured with the meridian circle at the Cape, while Herschel helped Maclear with his tidal observations. The latter were part of Maclear's official duties and Herschel was much interested in these measurements. His friend Whewell, who was working at the time on the mathematical foundations of general tidal theory, required for his computations a generous supply of observational material, which the two Cape Town astronomers were able to provide. The heights of sea level were measured at certain observing stations, and a number of markers from which the high tides could be read off at regular intervals were set up in Table Bay. This involved considerable labor, and since the financial support which Maclear's instructions empowered him to obtain from the Governor of the Cape Colony was insufficient, Herschel defrayed part of the costs out of his own

---

* The term fundamental stars is used to describe those stars whose positions on the celestial sphere are repeatedly measured as accurately as possible, with a transit instrument. They are important in the astronomical determination of place and time.

pocket. He also prepared interpolation tables and reduced the marker readings to a mathematically usable form.

Another field in which Maclear's and Herschel's interests overlapped was that of meteorological observations. At meetings of the South African Literary and Philosophical Institution, of which Herschel had been elected president only half a year after his arrival at Cape Town, a system of regular quarterly weather observations was agreed upon. These were to be made hourly on the days of the solstices and equinoxes. Despite his commitments in other directions, Herschel carried out this task until the end of his stay, thus helping to lay the foundations of systematic weather research in the Cape Colony.

Maclear's duties also included geodetic measurements, in which Herschel took only a minor part but supported by many suggestions of improvements in measuring technique. He himself determined the difference in longitude between Feldhausen and the Cape Observatory, and also the difference in altitude from barometric measurements. He suggested an astronomical method of extending the baseline for longitude measurements by observing the instant when particular localities on the surface of the moon became darkened in the course of a lunar eclipse.

Scarcely any branch of science escaped Herschel's attention during the four years at Feldhausen, though his interest was often sporadic and sometimes amateurish. Physical and chemical experiments are mentioned in his diary, his laboratory being a shed near the house which also contained the polishing equipment for his telescope mirrors.

Stimulated by the delightful and interesting vegetation of the environment of Feldhausen, Herschel discovered a predilection for botany. He planted a small garden near his house and cultivated rare and valuable plants collected on his expeditions. Most of these were tuberous and bulbous plants which make a colorful sea of flowers during the spring. He brought back to England an impressive collection of bulbs, most of which he presented to the Royal Horticultural Society for further cultivation.

The botanizing astronomer took especial pleasure in making drawings of the plants he had discovered and collected. For this purpose he used a camera lucida, a drawing instrument which was widely employed at the time and which he also used on his European trip with Babbage. This simple device, invented in 1807 by Herschel's friend Wollaston, consists of a doubly reflecting prism, or in Herschel's case two inclined sheets of plane glass. This is so arranged that the image of an object produced by it is seen by the eye as if lying on a sheet of paper placed beneath and the image can be traced out on the paper. The device, unlike the camera obscura (see Chapter 6), is used in full daylight, and Herschel's was a small pocket-sized folding arrangement. The instrument reproduces the object faithfully and to scale. Herschel filled a large quarto volume with drawings of plants, which were colored in masterly fashion by his wife.[23]

Herschel's drawings of typical Cape landscapes, including pictures of Feldhausen, are not artistic masterpieces, but they give a faithful historical account of the scenic background to his activities at the Cape. They were also executed with the help of the camera lucida.[24]

The picture of Herschel's many-faceted activities during those four years would not be complete without an account of his campaign in the field of public education, which provides an interesting sidelight on the outlook and personality of this unusual man.

Education in the Cape Colony during Herschel's time there was in a bad state. The enormous country, several times the size of England, was sparsely populated: there were some 60,000 Europeans and about twice that number of Africans, Malays, and Indians. Outside the few major cities and villages the population was thinly spread over a vast area with few roads. There were no schools except in Cape Town and a few other towns. Instead, itinerant teachers (known in Afrikaans as *Meesters*) traveled about the countryside giving very rudimentary lessons at odd places, in return for which they received from the farmers their food or sometimes a little pocket money.

In view of this unsatisfactory situation, John Bell, the Secretary to the Governor at the Cape, early in 1838 produced a memorandum for the attention of the British government. This had been preceded by discussions between the Governor, Sir George Thomas Napier, Bell, Herschel, and John Fairbairn, editor of a daily newspaper in Cape Town, and Bell's memorandum was submitted to Herschel for his opinion. Herschel expressed his views in a long letter, in which he starts with the premise that it is the duty of the state to set up and maintain a system of schooling and education in general. This idea was then a novel one; almost all the state schooling at the time consisted of some bare beginnings in Prussia and in two American states. (Herschel may have known about the report on the Prussian educational system prepared by the French philosopher Victor Cousin [1792–1867] on behalf of the French government, which appeared in English translation in 1834.)[25]

Herschel agreed with Bell that important preconditions for a successful educational system are centralized supervision by a government-appointed superintendent and a carefully worked out and consistent educational policy. He also demanded fundamental improvement in the social and economic position of teachers and the creation of a secure professional status. "To make the profession of education truly respectable it must be made an independent profession."[26] The government must offer adequate salaries and opportunities for promotion and initiative. Herschel's letter, with some proposals written by Fairbairn, was sent to the Colonial Office in Whitehall as an appendix to Bell's memorandum.

Herschel also wrote a memorandum setting forth his general ideas on education. In his view, the duties of a teacher were:

"1st to form in the individual advantageous personal habits.

"2nd to store his mind with useful knowledge and practical maxims available for the demands of active life.

"3rd to enlarge the powers and capacities of his mind and to elevate his propensities by familiarizing him with trains of connected and serious thought and with high examples of moral and intellectual conduct.

"4th to form good citizens and men by instructing them in the relations of civil and social life; and to fit them for a higher state of existence, by teaching them those which connect them with their Maker and Redeemer." [27]

These demands greatly transcended the educational principles of even the most civilized countries of Europe. Some of them have quite a modern ring, notably the idea of giving training in citizenship and the recommendation that schoolchildren should be active participants in the learning process. Herschel also recommended adult education in the form of public lectures of both specialized and general interest. He himself set a good example in this respect during his stay at the Cape by giving a series of popular lectures.

This letter to the government, setting out Herschel's fundamental ideas on the educational system, is dated May 23, 1839, a year after his return from the Cape. Presumably it was so timed in order to remind the Colonial Office of the recommendations in Bell's memorandum, which it had not by then taken any steps to implement. Thanks to Herschel's initiative, things now really started moving, and Herschel was informed after he had had a personal interview with the Colonial Secretary, Lord Glenelg, that the Government had accepted the plan and decided to appoint a superintendent and to send out to the Cape twelve teachers on Government salaries. Herschel's second proposal, which was to set up a secondary school "for giving the better conditioned classes a good English-like gentlemanly education" at Grahamstown, in the Eastern Cape Province,[28] was not implemented until many years later.

An impression has prevailed that very little is known about Herschel's private life in South Africa and Agnes M. Clerke confined herself to the laconic statement: "His family throve and multiplied at Feldhausen."[29] In fact, Herschel's diaries and correspondence during that period, published in 1969 in *Herschel at the Cape*,[30] contain a wealth of detail about his personal affairs and opinions and reveal the extraordinary qualities of his temperament and the breadth of his interests.

The Herschels arrived in South Africa with three children: William James, Caroline, and Isabella. Three more were born at Feldhausen: Margaret Louisa on September 10, 1834, Alexander Stewart on February 5, 1836, and John on October 29, 1837. The spacious, comfortable house of Feldhausen, with its large garden and innumerable nooks and crannies, provided an ideal home for the increasing family. Herschel's diaries do contain, juxtaposed with scientific notes, reports of merry family festivals and communal excursions into the surroundings of Cape Town. Sundays were usually days of contemplative rest and domesticity. When it was not possible to go to church in the neighboring village of Rondebosch, the Lesson and Psalms were read at home during the morning. The rest of the day was spent in walking, entertaining, or reading. Among the German classics, Herschel had an especial predilection for Schiller and later translated some of his poems into English. He also took delight in the poets of antiquity; in his old age he translated the *Iliad* into English hexameters (see Chapter 8). Herschel's lively interest in literature was not derived from his father. On the contrary, William Herschel once owned to his friend Dr. Charles Burney, the English musician, that he regarded poetry "as an arrangement of fine words without any adherence to the truth." [31] However, the father's musical talent seems to have descended to the son. John Herschel was a gifted and competent performer on the flute and violin. Since his wife also was musical, many nights not devoted to astronomy must have been spent by the couple at the music stand.

Herschel's literary and musical inclinations, his love of gardening, his skill in drawing, and his congenial family life all show his personality during the years at Feldhausen as that of a well-rounded, stable, and completely happy man, in contrast to the lonely concentration and turmoil of his life before his marriage. He himself wrote retrospectively of the period at the Cape: "Whatever the future may be and whatever the past has been, the days of our sojourn in that sunny land will stand marked with many a white stone as the happy part of my earthly pilgrimage." [32]

The atmosphere at Feldhausen exercised a peculiar charm over the many visitors who came to the hospitable household. Among these was Charles Darwin, whose voyage on the *Beagle* included a call at the Cape Colony and a visit to Herschel on June 15, 1836. Darwin had long wished to make the personal acquaintance of the astronomer, for whom he had great respect. As is often the case at such meetings, he found his host to be quite different from what he had imagined. He describes him as a very modest man, rather shy and even gauche in society, despite his lively intellect. This modesty of Herschel's seems to have made a particularly deep impression on Darwin.

Herschel's diffidence extended beyond his relationships with people. It appeared in the exposition of his scientific achievements and sometimes even in their mere announcement. This circumstance may have contributed to one of the more comical episodes in the history of science, which caused a considerable stir at the time, and in which Herschel's name became involved without his knowledge.

An enterprising contributor to the American daily newspaper the New York *Sun* in the summer of 1835 had the idea of using Herschel's astronomical investigations at the Cape as the basis for a series of articles containing reports of fantastic discoveries of objects on the surface of the moon, supposedly made by Herschel with the aid of a completely new type of giant telescope. After a skillfully constructed introduction, adorned with scientific arguments, describing the principles and construction of this miraculous telescope, the reader was regaled with a flowery description of the moon, its paradise-like woods and meadows, hills and valleys, and even its living organisms, the whole containing every conceivable fantasy. The series aroused much interest, and the circulation of the *Sun* soared overnight. With the slowness of communications in those days, many weeks passed before Herschel heard anything about the articles, which of course were a fabrication from beginning to end. Still more weeks passed before his denial was published. In the meantime the story had spread to Europe. Soon Herschel's "moon men" were the main topic of conversation in all the literary salons and cafés. A publishing house in Hamburg even served up the moon hoax in book form with an authoritative-sounding title.[33] That the innocent victim of such a trick should have been Herschel, to whom truth was always the highest goal of his scientific efforts, lends a tragicomic note to the entire affair.

Toward the end of 1837 Herschel's astronomical activities at the Cape were reaching their conclusion. The goal had been achieved; the great task of completing and rounding out his father's work had been done. In barely four years he had made a series of investigations whose extent and importance were great enough for a lifetime's work. The years at the Cape of Good Hope were not only the happiest of his life, they also marked the brilliant climax of his astronomical achievements and the conclusion of his astronomical career. The great reflecting telescope was never used after his return to England, but lay rotting, with a tarnished mirror, in the cellar of his country house, Collingwood, in Kent. It was resurrected only about 1958 to be set up by the National Maritime Museum in the old Royal Observatory at Greenwich as a memorial to its illustrious builders and owners.

While the statement, sometimes made, that Herschel never looked through a telescope again for the rest of his life is not strictly true, his activity as a research astronomer and practical observer of the heavens had essentially come to an end. He was only forty-eight years old. His father at forty-six had just begun to devote himself intensively to astronomy and he made his greatest discoveries in his fifties. John Herschel seems to have deliberately exchanged the telescope for the writing desk, from which he enriched astronomy with major books and encyclopedic articles. There were probably two main reasons for this. One was that his health was no longer vigorous enough to endure the physical exertion of active observing; the other was that he wished to devote his whole energy to the publication of the results of his South African researches. This task was immensely time-consuming, since he carried out all the reductions of his observations himself. In a letter to a scientific friend in April 1839, he wrote: ". . . with the publication of my South African observations (when it shall please God that shall happen) I have made up my mind to consider my astronomical career as terminated."[34]

On March 11, 1838, Herschel embarked with his family on the *Windsor* for the voyage home, and the African adventure was over. A stone obelisk, erected by friends on the site at Feldhausen where the great reflector had stood, still stands to commemorate his exploration of the southern celestial hemisphere.

5

Constellation Reform

and Terrestrial Magnetism

✦

✦

✦

In these days, it is hard to realize that in Herschel's time a voyage from Cape Town to London was a daring and dangerous undertaking, for which a happy ending was not by any means a matter of course. An idea of the annoyances associated with such long voyages in sailing ships is provided by an entry in Herschel's diary in April 1838: "To live on board ship is equivalent to living in a perpetual earthquake on land with the prospect of being drowned instead of crushed, as a catastrophe." [1]

The *Windsor* was a stable ship; the usual storms of the Bay of Biscay, which lay on the route, did not do it much harm. The voyage, while less comfortable than the trip out, on the whole passed off well, and provided opportunities for astronomical observation, studies of marine fauna, and camera lucida drawings at the island of Saint Helena, where the ship called. The *Windsor* docked at the end of May, after nine weeks, delayed by a contrary northeast wind, but without serious incident, and Herschel trod his native soil again after an absence of four and a half years.

His diary, laconic at the best of times, contains only a few entries for the weeks of upheaval and turmoil immediately following his return. The baggage had to be unpacked and tidied away, letters had to be written, calls paid, and interrupted contacts resumed.

Herschel was so showered with congratulations and honors on the successful conclusion of his enterprise that for the time being he was left with little peace to resume his scientific work.

One of his first visits was to Lord Glenelg, the Secretary of State for the Colonies, to whom the reports on the colonial educational system had been directed. Although the diary states only: "Dined at Lord Glenelg's," [2] it can be assumed that Herschel's plans for the reformation of the educational system in the Cape Colony were discussed in detail.

Outstanding among the honors and distinctions received by Herschel during the ensuing months was his elevation to the rank of baronet, which took place on the occasion of the coronation of Queen Victoria in Westminster Abbey on June 28, 1838. When Herschel was first offered this honor by the Duke of Sussex, he wished to decline, on the characteristic grounds that his retired mode of existence was incompatible with public honors and the obligations of such a position would interfere with his scientific researches. The duke, however, succeeded in persuading him to accept the title.

A few days after the ceremony, Herschel set off with his five-year-old son William James on a journey to Hanover, where Caroline Herschel, despite her advanced age, still maintained physical and mental fitness.

A detour to Göttingen led to a meeting with Gauss, and Herschel visited the German astronomer Heinrich Wilhelm Matthäus Olbers (1758–1840) at Bremen. The latter, who was self-taught in astronomy, being a medical doctor by training, had made a considerable name for himself by his discoveries of comets and asteroids. He had also discovered the genius of Bessel at an early stage and had done everything he could to advance his career. Olbers had transferred his friendship from William Herschel to his son, who had the highest regard for his father's old friend. Herschel also stopped in Hamburg for a visit to the German-Danish astronomer Heinrich Christian Schumacher (1780–1850), editor of the *Astronomische Nachrichten*.

Returning to London on August 4, Herschel was soon deep in scientific activities. He had discussions with his friends Francis Baily and the Astronomer Royal George Biddell Airy (1801–1892) on the problem of reforming the constellations and fixing their boundaries. So many new constellations had been introduced since the production of star charts by the German astronomers Johann Bayer (1572–1625) and Johannes Hevelius (1611–1687) that the subject was now in a state of thorough confusion. Confusion also reigned in the southern celestial hemisphere, which had been divided by Lacaille into numerous constellations often very difficult to identify. Herschel returned to the topic several times, and, in the *Memoirs of the Royal Astronomical Society* for 1841 he proposed a plan, worked out in collaboration with Whewell and Baily, for the complete rearrangement of the constellations in the southern sky.[3] It was a sensible proposal, which involved replacing the often ill-defined constellations, delineated quite arbitrarily on the older charts, by a clear division of the sky into separate regions. The more striking star groups would be combined into new figures bounded by spherical quadrangles composed of hour and declination circles. In existing charts and catalogues certain stars appeared sometimes in one constellation and sometimes in another, while other stars frequently belonged to none but were abandoned in a sort of celestial no-man's-land. In Herschel's scheme, all stars included in the quadrangle surrounding the constellation of Orion, for example, were to be placed in the Regio Orionis. The corners of the regions, Herschel proposed, should be treated as "artificial stars" having definite right ascensions and declinations reduced to the equinox of the relevant celestial chart, so as to take account of the shift in coordinates caused by precession. This would prevent the constellations from gradually drifting across and out of their regions in the course of time, thus making the divisions illusory.

Herschel's plan was not accepted by astronomers on the continent, and he withdrew it. However, it anticipated the method that is now universally adopted in celestial cartography. The boundaries of the constellations were fixed at a meeting of the International Astronomical Union in 1928 and were later adopted in all civilized countries. The arrangement is based on the same principle as Herschel's; the only difference is that the boundaries, instead of being spherical quadrangles, are irregular many-sided figures arranged to keep certain stars within their traditional constellations. The modern boundaries, like Herschel's, follow meridians and parallels.

In 1844, Herschel presented a revised proposal for reform to the British Association for the Advancement of Science, to which he belonged as a member of various of its commissions.[4] This proposal had already been followed by Baily in his reduction of Lalande's great catalogue of nearly 50,000 stars,[5] and also in the reduction of Lacaille's catalogue, to which Herschel contributed the preface.[6] In general, Herschel's proposals preserve the southern constellations marked out by Lacaille, but in certain cases Lacaille's nomenclature is abandoned—for example, in the very large and not easily recognizable constellation Argo, which Herschel divided into four parts. His other suggestions are chiefly concerned with the simplification of names, the choice of letters for individual stars, and similar details.

The reform of constellations and nomenclature was

only one of the tasks that awaited Herschel on his return from the Cape. In August 1838 he traveled to Newcastle for the annual meeting of the British Association, where, very much against his wishes, a triumphal reception had been prepared for him. He was the hero of the meeting; he was elected president of the Mathematical Section and loaded with duties. He was also elected to the Geomagnetism and Meteorology Committee and thus became involved in a project that required all his energy and attention.

At Cape Town, as mentioned in Chapter 4, Herschel had developed a scheme for hourly meteorological observations on selected days of the year.[7] The system had been tested successfully in the Cape Colony and was to be copied throughout the British possessions. The Royal Society had also formed a committee concerned with meteorology, to which Herschel regularly communicated his reports and papers. In the development of geomagnetic research in England, Herschel also made important contributions; it was largely owing to him that, within a few years, a number of geomagnetic stations were established in England and in the Colonies. Herschel suggested that these be combined with weather stations and be called "Physical Observatories."

The first initiative for carrying out this kind of research had been taken by Alexander von Humboldt, who had approached the Royal Society in 1836 with a request that it set up magnetic observatories in the British possession. This proposal had been readily accepted, and the British Government had undertaken to meet the costs of construction, instrumental equipment, and manning the stations. The British astronomer and surveyor Edward Sabine (1788–1883) was entrusted with the organization of the enterprise. In 1838, Herschel, Whewell, Airy, and the physicist Humphrey Lloyd (1800–1881) joined Sabine and energetically supported his efforts. Thanks to Herschel's scientific prestige, now at its height, the plan was implemented in a remarkably short time.

At the annual meeting of the British Association at Plymouth in 1841, Herschel gave a report which showed that a considerable amount had been done within three years.[8] In the British Colonies there were now geomagnetic stations at St. Helena (established August 1840), Van Diemen's Land (now Tasmania; established October 1840), Cape Town (established March 1841), Madras (established January 1841), and a few other places. These stations sent regular observations to the Royal Society, to which were added the results of numerous other observatories in Germany, France, Belgium, Italy, Norway, and other European and non-European countries. Herschel's report mentions fifty-one geomagnetic stations that had by 1841 either begun regular work or were about to start research programs. The most important working instrument at these observatories was the magnetometer which had been developed at Göttingen by Gauss and the German physicist Wilhelm Eduard Weber (1804–1891). On the occasion of the award of the Copley Medal to Gauss in 1838, Herschel wrote to him: "What may be the success of these applications, it is impossible yet to say, but I have great hopes that something will be done worthy of this country."[9]

The analysis of the observational material resulting from these measurements was done, not by Herschel, but by Sabine, who ten years later pointed out the correlation between disturbances of the compass needle and events in the solar atmosphere, after the discovery of the eleven-year periodicity of sunspot numbers by the German astronomer Heinrich Schwabe (1789–1875) and that of a similar periodicity in fluctuations of the earth's magnetic field by the Scottish-German astronomer Johann von Lamont (1805–1879) at the Munich Observatory.

Herschel's role in the development of research in geomagnetism was essentially organizational and advisory. In a letter to Lloyd dated October 24, 1842, he stated frankly that he had never made a geomagnetic observation in his life and had offered his collaboration only because he thought that his general experience in carrying through scientific projects and his access to influential persons in the scientific and administrative spheres would be helpful in a matter that he considered "worthy of national support."

Moreover, Herschel played a major part in the organization of a program of geomagnetic research for the South Polar Expedition conducted by the Scottish explorer Captain James Clark Ross under the auspices of the Admiralty. The first initiative for this enterprise had come, once again, from Humboldt, who had been stimulated by Gauss's researches to take a lively interest in terrestrial magnetism.

James Ross had accompanied his uncle Sir John Ross on a voyage to the north of Canada in the period from 1829 to 1833 and on that occasion had discovered the North Magnetic Pole on the peninsula of Boothia Felix at latitude 70° north. His forthcoming voyage to Antarctica (1839 to 1843) was intended to take him to the opposite Pole. With his two ships *Erebus* and *Terror*, he set out southeastward from Tasmania and discovered Victoria Land and a nearby island (afterward christened Ross Island). He was blocked by an impenetrable barrier of ice from reaching the South Magnetic Pole, but measurements with the magnetometer showed that the Pole must be farther north and somewhat inland; its exact position was computed by Ross to be 75° 5′ south latitude and 151° 45′ east longitude. Ross's two expeditions thus led to the discovery of both magnetic poles.

Herschel's share in the planning of the Antarctic expedition cannot now be ascertained in detail, but various entries in his diary and brief remarks in letters to his scientific correspondents show that he took a lively and active part. In the letter to Gauss already mentioned, he wrote that it was planned "to send an expedition for the purpose of magnetic observations in the highest attainable southern latitudes." [10] In October 1838 he noted in his diary: "Dined today with the Queen at Windsor Castle where had much conversation with Lord Melbourne about the projected South Polar Expedition,"[11] and on November 6: "Went up to town to confer with Sabine and Ross and Beaufort on the magnetic expedition." [12]

After final discussions with Sabine and Peacock, Herschel presented to the prime minister, Lord Melbourne, on November 11 a memorandum which seems to have contained the proposals of the Magnetic Committee of the Royal Society and of the British Association, to which Herschel belonged. On March 11, 1839, he received from Lord Minto, First Lord of the Admiralty, the government's official approval to mounting the expedition along the lines recommended, with the intention that it should be ready to sail by June of that year.

After the return of the Antarctic expedition, Herschel wrote a preliminary report for the annual meeting of the British Association at York in 1844, in which he summarized the most important results.[13] This seems to have concluded his active participation in geomagnetic research.

In his eagerness to find time for the reductions of his South African observations, Herschel had carried on numerous official scientific duties imposed upon him after his return from the Cape with a heavy heart and only after much hesitation. The various honors conferred on him, no doubt with the best of intentions, made him a national scientific hero, but did not give him unalloyed pleasure. He felt them to be a burden which he tried to shake off as well as he could.

He was very unhappy, therefore, when a move was

made to elect him to the highest position in British science, the presidency of the Royal Society, after the retirement of the Duke of Sussex in 1838. Even before he was approached with an official offer, Herschel implored Whewell to explain in no uncertain terms that it was impossible for him to take up the office because it would lead to a sacrifice of his "most cherished scientific projects and domestic arrangements"[14]; he hoped, therefore, that his name would not be mentioned in connection with the matter in future. He made a similar request to Baily.

From another office, however, he could not easily stand aside and probably did not wish to do so. This was the presidency of the Royal Astronomical Society, with which he was closely connected as one of its founders and a previous president. He was elected for a second time in 1839 and held the office from 1839 to 1841.

In 1842 he accepted a purely honorific post involving no duties; the Marischal College of Aberdeen elected him its Lord Rector.

On June 12, 1839, the University of Oxford conferred on Herschel (along with the poet William Wordsworth) the honorary degree of Doctor of Civil Laws. However, Herschel declined Oxford's offer of the Savilian Professorship of astronomy, and a further invitation to represent the University of Cambridge in Parliament. From 1843 onward he was a member of the Board of Management of the British Museum, and also of the Board of Visitors of the Royal Observatory, which made an annual inspection of the Observatory at Greenwich.

Academies and learned societies in nearly all the countries of Europe and several on other continents did themselves honor by including Herschel among their corresponding members. It would be tedious to name them all and indeed it seems doubtful whether Herschel himself remembered all his various honorary posts and memberships. The long list includes the award of the Civil Division (Friedensklasse) of the highest Prussian award, Pour le Mérite (1842), and the distinction of being among the eight foreign Associates of the Institut Français.

It was not easy for Herschel to disengage himself from the shower of honors and return to a life of privacy and seclusion. This was one of his reasons for purchasing in 1840 the fine, somewhat isolated country house in Kent where he spent the remainder of his life. The new home provided the tranquillity that he needed to carry his life's work to completion. During 1939, however, new interests arose.

6

Photography and Photochemistry

✦
    ✦
✦

On January 22, 1839, Herschel received from his friend Sir Francis Beaufort (1774–1857), Hydrographer of the Royal Navy, a letter reporting an interesting invention. A Frenchman named Louis Jacques Mandé Daguerre (1789–1851) had developed a process [1] for producing pictures of the greatest clarity and precision by exposing certain photosensitive silver salts to light in a camera obscura. This was an optical device then widely used as an aid to drawing, in which an image of a scene was projected (usually by a suitable lens) onto a sheet of paper in a dark room. The device resembled a modern camera in many respects but usually on a much larger scale.

Herschel, always ready to be enthusiastic about a new discovery, was promptly stimulated to undertake similar experiments in his small domestic laboratory at Slough. He recorded his work in a notebook entitled "Chemical Experiments":

"Slough, Jan. 29. Effects tried within the last few days since hearing of Daguerre's secret and that Fox Talbot has also got something of the same kind . . . Daguerre's process attempted to imitate. Three requisites:

    1. very susceptible paper

    2. very perfect camera

    3. means of arresting further action

Tried hyposulphite of soda [sodium thiosulphate] to arrest the action of light, by washing away all the chloride of silver or other silvery salt. Succeeds perfectly—papers ½ acted on ½ guarded from light by . . . pasteboard were when withdrawn from sunshine sponged once with hyposulphite [of] soda then well washed in pure water—dried and again exposed. The darkened ½ remained dark, the white half white after any exposure. . . ."[2]

To understand the significance of these notes, a summary of the work that had preceded Herschel's experiments is needed.

The process invented by Daguerre was as follows: First a silvered copper plate was exposed to iodine vapor to form silver iodide. The plate was then placed in a camera obscura at the focus of the camera lens and exposed to light from a few minutes to a quarter of an hour. It was then removed and treated with mercury vapor. The mercury was immediately deposited on those places on which light had fallen, and a bright and very sharp picture with a metallic sheen was produced.

Daguerre was not the first to attempt to produce "photographs" by the use of chemical substances. The sensitivity of silver salts to light had been known for a long time. Some isolated experiments were made during the eighteeenth century, and at the beginning of the nineteenth more serious attempts were made by Thomas Wedgwood (1771–1805) and Humphry Davy. Wedgwood, who was the youngest son of the founder of the English china industry, studied chemistry at Edinburgh and, encouraged by the English chemist Joseph Priestley (1733–1804), devoted himself to the study of the chemical effects of light and of the relationships between light and heat. He was a close friend of the somewhat younger Humphry Davy, who was appointed director of the chemical laboratory of the Royal Institution in 1801, was president of the Royal Society from 1820 to 1827, and was generally regarded as the leading chemist of his day.

At the beginning of 1802, Davy and Wedgwood repeated a series of experiments, begun earlier, in which they tried to produce photographic pictures by the action of light on paper and leather coated with silver-nitrate solution. They actually succeeded in reproducing paintings and drawings. These experiments lasted only a short time. Wedgwood died in 1805, and Davy turned to other problems. They had not succeeded in finding a method of preventing the gradual destruction of the images by the further action of light. In other words, they had not succeeded in "fixing" their pictures by making them insensitive to light.

Seventeen years later, in 1819, Herschel discovered that sodium thiosulphate had the property of dissolving silver salts and, as mentioned earlier, announced his discovery in the Edinburgh *Philosophical Journal*. Although Herschel was in close scientific contact with Davy (he was secretary of the Royal Society while Davy was president), he seems to have known nothing about Davy's and Wedgwood's photographic experiments or the reason for their failure. Otherwise, he surely would have drawn Davy's attention to this property of sodium thiosulphate, thus providing him with the indispensable fixing agent for his photographs. Davy, on the other hand, seems to have overlooked Herschel's discovery, although he might have been expected to take a lively interest in the compounds of sodium which he himself had discovered.

The experiments of Davy and Wedgwood were apparently forgotten, and for more than thirty years nothing similar was done in England, except perhaps by Thomas Young.

Things were different in France, where the physicist and amateur lithographer Joseph Nicéphore Niepce (1765–1833) carried out numerous experiments with photosensitive substances, though for some time without success. Finally Niepce conceived the idea of coating a glass plate with a thin sheet of asphalt that had previously been dissolved in petroleum. Asphalt, a substance widely used in lithography because of its resistance to acids, has the property of being partly bleached by light and also hardened, while the portions unexposed remain soft and soluble. By laying a transparent drawing over the plate Niepce succeeded in reproducing a print of it. Then he dissolved away the portions of the asphalt layer that had not been hardened by light by washing the plate with a mixture of oil of lavender and petroleum and thus obtained a permanently fixed image. He called this process "heliography."

On December 14, 1829, Niepce entered into an agreement with his fellow countryman Daguerre, who had also been experimenting in the same field for years. Niepce died only four years later without having enjoyed the fruits of his labors, which Daguerre now began to collect. After a series of experiments with iodized silver plates, Daguerre reached his goal by means of a chance discovery. He found that a sharp image with a bright, metallic sheen was produced by mercury vapor, because mercury was deposited on the exposed portions of the plate. To fix his image, he at first used solutions of common salt; later he used sodium thiosulphate, after Herschel had described it as a solvent for silver halides.

A short announcement of Daguerre's process was made to the Académie des Sciences in January 1839, and a description published by Arago in August, thus bringing Daguerre's work to wider public attention. The short exposure time required for the "Daguerreotypes," as they soon came to be called, was cited as a particular advantage over Niepce's asphalt process. Niepce had had to expose his plates for several hours, but Daguerre's highly photosensitive silver iodide gave clear pictures after only a few minutes. In his enthusiasm, Arago seems to have overlooked the fact that Niepce had laid the foundations on which Daguerre had built.

In England, serious efforts to produce photographic images had first been resumed in 1834. While sketching landscapes with the camera lucida, William Henry Fox Talbot (1800–1877), a mathematician and Fellow of the Royal Society, began looking for ways to impress his pictures permanently on paper by other means than a pencil. A poor draftsman, he wished to fix camera-obscura images to produce accurate pictures of nature. From the chemical literature, he knew of the photosensitivity of silver salts. He took fine-grained writing paper with the smoothest surface available, soaked it in a solution of common salt, dried it, and then soaked it in silver-nitrate solution. To enhance the sensitivity of the silver layer enough to make the feeble illumination in a camera obscura adequate to produce images, he coated his paper with salt and silver-nitrate solutions several times and exposed the wet sheets. After some unsuccessful attempts, he designed and had constructed a small camera with a short-focus microscope objective. With this, he succeeded in producing a picture of a window of his house, Lacock Abbey, in August 1835.

On January 31, 1839, Talbot presented a short paper on his process to the Royal Society and showed a number of photographs. Most of these attempts were reproductions of drawings, prints, leaves of plants, and the like, which he had impressed on sensitive silver-chloride paper by the action of direct sunlight. There were a few photographs of scenes taken with the camera obscura, among them the picture of Lacock Abbey. The paper was entitled "Some account of the art of photogenic drawing," [3] the name he had given to his process. In a further note on February 21, 1839, Talbot described the chemical and technical details of his invention.

Talbot's "photogenic drawings" seem to have made very little impression at first. They were completely overshadowed by the brilliant success of the daguerreotypes, which required much shorter exposure times and produced sharper images. Nevertheless, there is something to be said in favor of "Talbotypes," as they were soon christened by their supporters. Talbot can fairly be regarded as the founder of modern photocopying processes. He pointed out that light and shade can be reversed by placing an illuminated, fixed picture, made translucent by wax, on unexposed sensitized paper. In this way all the illuminated (blackened) details of the original picture come out bright on the copy in accordance with their original appearance, and conversely all unilluminated (bright) portions come out dark.

Talbot was not discouraged by initial lack of success. On February 8, 1841, he patented an improved version of his process, in which fine-grained paper was treated alternately with solutions of silver nitrate and potassium iodide, then washed with silver-nitrate solution containing tannic acid, and afterward exposed to light in the camera. After a few minutes' exposure the paper was washed for a second time in tannic-acid solution, whereupon the initially invisible "latent" image was gradually developed. This process, which was similar to Daguerre's use of mercury vapor, was accelerated by careful heating. As a fixer, Talbot, though he was acquainted with Herschel's sodium thiosulphate, first used potassium-bromide solution; later he used sodium thiosulphate and included it in a patent which he took out in 1843.

Talbot called the products of his new and much more rapid process "calotypes" (Greek καλοσ, beautiful) because the pictures took on an attractive, warm color tone after development and fixing.

Another English experimenter with photography during this period was a clergyman named J. B. Reade (1801–1870), who continued the earlier work of Wedgwood and Davy with paper and tanned leather. He realized that the tannic acid in leather increases its sensitivity to light enormously and with the aid of a solar microscope he succeeded in producing photographic images on paper, which he called "solar mezzotints." Reade was also indisputably the first person to recognize the significance of Herschel's sodium thiosulphate as a fixing agent and he applied it even before Herschel recommended it to Talbot. He had seen it mentioned in the *Manual of Chemistry* by the English chemist William Thomas Brande (1788–1866).

Herschel recognized the fundamental significance of his own discovery that sodium thiosulphate dissolves silver salts only when he heard of the attempts of Daguerre and Talbot and began to make photographic experiments on his own account. He realized that sodium thiosulphate was an ideal fixer in all photographic processes involving silver halides, for by its use the silver salts not exposed to light could easily be washed away, thus preventing any further blackening of these areas and hence a gradual destruction of the picture. This was a great advance over the fixing methods of Daguerre and Talbot, since sodium thiosulphate not only provided completely permanent fixing but also improved the quality of the picture.

It never occurred to Herschel to try to derive any personal advantage from this important discovery—for example, by taking out a patent. He told Talbot about the method, when the latter paid a visit to Slough on February 1, 1839, and Talbot, as has been mentioned, included it in one of his own patents, after he became convinced of the inadequacy of his own fixing agents. On February 9, 1839, Herschel had written Talbot that he not only had no objection to sodium thiosulphate being mentioned as a fixer but even wished Talbot to make it known quite freely, "either publicly or privately." Later, Talbot openly admitted that Herschel's fixing method was vastly superior to potassium bromide, although for a while he had obstinately adhered to the latter.

Daguerre, on the other hand, immediately after the publication in March 1839 of a short paper by Herschel,

"Note on the art of photography . . . ,"[4] dropped his process involving common salt solution and incorporated sodium thiosulphate into the description of his patent.

Herschel's paper was illustrated by twenty-three pictures, most of which he had made by placing drawings and prints on sensitive paper and afterward fixing them. One, however, was a picture made with a camera obscura on January 30, 1839. Herschel had placed in his camera a sheet of paper sensitized with silver-carbonate solution, and after two hours' exposure produced a sharp and clear picture of the 40-foot telescope in his garden at Slough.

While it seems remarkable that Herschel, in a single week, working entirely from his own resources, was able to reproduce and improve upon an invention on which other people had been working for several years, and about which he knew very little, the explanation is probably simple. Among the first pioneers in photography, Herschel was the only one who was a competent chemist with both theoretical knowledge and practical experience, and consequently a thorough acquaintance with the relevant chemicals. The knowledge that Daguerre, Niepce, and Reade had to discover by endless experiments of a purely empirical kind, or else came across by sheer good luck, was already established in Herschel's mind and provided a foundation on which he could build. Thus, immediately upon hearing the news of Daguerre's invention, he could write in his diary: ". . . a variety of processes at once presented themselves,"[5] a situation which his contemporaries who were experimenting in photography might envy. The fact that he was also a physicist, with brilliant qualifications in theoretical and practical optics, gave him a further advantage.

Herschel's photographic process started with paper that had been sensitized to light by coating it with silver-carbonate solution. Furthermore, he recognized that only an objective lens that was entirely free from spherical or chromatic aberration, so as to give a completely sharp focus, could give perfect pictures, and that the focal plane needed to be evenly illuminated. His process was completed by the use of sodium thiosulphate as a fixing agent.

Herschel recognized the essentials of the photographic process and the conditions needed for the production of photographs without any influence from other people's work. He used a different method from that of Daguerre, although he was not immediately aware of this. He wrote in his Chemical Notebook on January 30, 1839: "Thus Daguerre's problem is so far solved." [6]

Even after Herschel learned that his process was different from those of Daguerre and Talbot, he never claimed any rights as an inventor. On the contrary, he asked that the article on photography which he had sent to the Royal Society on March 14, 1839, be withdrawn from publication in the *Philosophical Transactions* and only a "note" be published in the Society's *Proceedings*. He explained this step on the grounds that he did not wish to interfere with "Mr. Talbot's just and long antecedent claims." [7]

Talbot, on the other hand, was less generous. During his visit to Slough, Herschel freely explained his whole process to him, whereas Talbot ". . . did not explain his process of what he calls 'fixing,'" as Herschel remarked in his notebook. [8]

Before the various special processes and discoveries which Herschel contributed to photography can be described, some discussion of nomenclature is required. There has been much controversy over the invention of the term "photography." According to Harold White, the word was used as early as February 2, 1839, in a letter to Talbot written by the English physicist and inventor Sir Charles Wheatstone (1802–1875). [9] On February 10, Herschel appended to one of his photographs of the great reflector the abbreviated caption "photogr." In his diary the first use of the word appears in the entry for February

13, 1839.[10] It occurs repeatedly in the entries for the following days, and in a letter to Talbot dated February 28, 1839, Herschel recommends the use of the adjective "photographic" instead of "photogenic." The word was first used officially by Herschel on March 14, 1839, when he submitted the article already mentioned to the Royal Society. This paper was long held to be the baptismal certificate, as it were, of the new art, but in 1932 Erich Stenger discovered an article in the *Vossische Zeitung* for February 25, 1839, by the German astronomer Johann Heinrich von Mädler (1794–1874), in which the word *Photographie* was used.[11]

While Herschel cannot be regarded as the sole creator of the word "photography," he did define two terms which rapidly entered into common technical usage in the field. In a paper published in 1840, he wrote: "To avoid much circumlocution, it may be allowed me to employ the terms positive and negative, to express respectively, pictures in which the lights and shades are as in nature, or as in the original model, and in which they are the opposite, i.e. light representing shade and shade light."[12]

The photographic process begins by blackening the silver layer to produce a negative picture, which is afterward changed into a positive by means of a further exposure on photosensitive paper. An isolated exception to this was the daguerreotype, in which the deposition of mercury on the exposed portions of the plate led directly to the appearance of a positive picture.

Herschel also tried to produce direct positives on sensitized paper. He discovered that if silver-chloride paper is blackened by exposure to sunlight and coated with potassium-iodide solution before a further exposure in the camera, a positive picture is produced because the second exposure leads to bleaching instead of further blackening. This method had also been developed by others, including the French chemist Jean-Louis Lassaigne (1800–1859), at about the same time, so that it is difficult to assign priorities. However, Herschel's attempts to produce positives in a direct process are known to have begun with his earliest attempts at photography. He referred in his chemical notebook on February 1, 1839, to "Trials to reverse the colours, or to make a black picture on white ground *without* any double transfer." [13]

Herschel also made experiments with sensitized glass plates. In Talbot's paper pictures, the roughness of the paper had a bad effect on the sharpness of the images, and furthermore the organic materials of the paper were damaged by the chemical reactions induced by light. Both of these disadvantages were eliminated by the use of glass plates. A further advantage was the transparency of the picture on glass, as a result of which one could make any number of positive prints without further processing, simply by placing the negative on sensitized paper. Herschel also devised a still simpler method of making an apparent positive by blackening the glass negative on the back or fastening it on a dark background.

Herschel's glass-plate process was as follows: The plate was thoroughly cleaned and placed in a shallow dish, then a highly diluted solution of common salt and silver nitrate was poured over it. Finely divided silver chloride was formed and deposited in a thin, uniform layer on the plate. Herschel then sucked away the excess solution with a pipette and let the silver-chloride layer dry. Shortly before exposing the plate in the camera he added more silver-nitrate solution, carefully rocking the plate to and fro, and exposed it wet.

In the 1840 paper already mentioned, he wrote: "Exposed in this state to the focus of a camera with the glass towards the incident light, it became impressed with a remarkably well-defined negative picture, which was direct, or reversed, according as looked at from the front or the back." [14] When the plate was treated with a solution of sodium thiosulphate, the picture at first disappeared, but after the plate had dried it reappeared and was permanently fixed.

The subject of this photograph was the 40-foot telescope at Slough, and the picture was taken September 9, 1839, shortly before the telescope was dismantled. Hence the first photograph taken on a glass slide has documentary interest in the history of astronomy as well as photography.

Herschel photographed the telescope several times, using various silver salts for the purpose. He discovered that silver bromide had much the greatest sensitivity to light and obtained a clear picture after only a few seconds' exposure. The fixing was so perfect that half a century later, in 1890, his son Alexander Stewart Herschel was able to produce twenty-five paper prints in an enlarging machine from the original plate, preserved at the Science Museum in London.

Silver bromide had already been used by Talbot in his paper process, but apparently without much success; it was Herschel who pointed out the immense significance of this salt in photography. He once wrote of the need for "a new photography to be created, of which bromide is the basis." [15]

Herschel found from innumerable experiments that a great variety of organic and inorganic substances are sensitive to light. However, the practical utility of these substances for photography was variable, and most of the processes are now of historical interest only, although they undoubtedly contributed to the progress of the art.

The "cyanotypes" are an exception, in that they have had practical importance. A detailed description of this process is contained in a paper which Herschel presented to the Royal Society in June 1842.[16] He gave his photographic paper a coating of ferric ammonium citrate. By overlaying transparent drawings, prints, or manuscripts, he obtained after half an hour's exposure a latent image which he developed by treatment with a solution of yellow potassium ferrocyanate into a positive picture of purple color and extraordinary sharpness. Washing with water produced a fixed picture of bright Prussian blue—a "blueprint." The final picture lost some of its sharpness owing to the porosity of the paper, but this effect could be arrested to some extent by treatment with a solution of gum arabic.

Experimenting with organic salts of iron, Herschel discovered that when ferric ammonium citrate is reduced to the ferrous salt by the chemical action of light, it forms metallic deposits if the paper is washed with gold chloride solution after exposure. He called this gold-salt process the "chrysotype." The use of iron salts also led to the "amphitype," which could be made into a positive or a negative at will. This process, announced at the meeting of the British Association at York in 1844,[17] consisted of treating the paper with iron citrate, iron tartrate, mercury oxide, lead suboxide, or the corresponding nitrates and then placing it in a bath of ferric ammonium citrate solution. The exposure time varied from half an hour to between five and six hours, according to the intensity of illumination. A velvety brown negative was produced, which seemed to disappear in a short time; actually the image remained latent. A positive could be produced at any time by immersing the negative in a watery solution of mercuric nitrate and washing with warm water. First a faint positive of bright yellow color appeared, and when this was dried, a deep black positive picture was produced.

Another area of Herschel's research with even less application to practical photography was his experiments on the effects of sunlight, dispersed by a prism, on various salts and vegetable dyes. These involved exposures of sometimes up to several weeks and were generally not susceptible to permanent fixing. However, the experiments were of very great scientific interest because of the immense variety of effects that could occur under different physical and chemical conditions. That these researches were fundamentally different from the photographic experiments already described is indicated by the titles of his first publications in this area. "On the chemical action of the rays of the solar spectrum on preparations of silver and other substances, both metallic and non-metallic . . ." appeared in the *Philosophical Transactions* for 1840 [18] and was followed in 1842 by "On the action of the rays of the solar spectrum on vegetable colours and on some new photographic processes." [19]

Herschel was now primarily concerned with problems in pure science. Though he retained his interest in the practical aspects of photography and invented many processes during this period and later, the mere production of photographs lost its importance for him after he had succeeded in producing and fixing his own pictures.

It was inevitable that Herschel's researches should lead to new practical applications to the photographic process of the photochemical reactions of the substances he used. The wealth of his contributions to photography in its early days proves that he exploited these possibilities to the full. However, his contributions may have been the result of the pleasure that he characteristically took in experimenting for its own sake. His deeper scientific interest had always been directed at the causes of the complex interaction between physical and chemical forces that Talbot picturesquely called "the Pencil of Nature," rather than at its effects.

This fact gives Herschel's contributions to the early development of photography an importance which has not been appreciated by most of its historians, who usually mention his work in passing, if at all. Herschel was not merely one of the host of simultaneous inventors of the photographic art, nor was his contribution confined to the discovery of a few new processes or the enrichment of the technique by some subtle devices. Among the research on photography from the scientific point of view, Herschel's work on the chemical reactions induced in silver salts and vegetable dyes by light from various regions of the spectrum was an early contribution to photochemistry. This branch of science was relatively neglected until the beginning of the present century but is now an important branch of physical chemistry. Photography itself has benefited greatly from knowledge provided by photochemistry; although the photographic pioneers used mostly empirical methods, modern photography with its broad applications and high degree of refinement would be impossible without the scientific basis of photochemistry.

Herschel's essays in photochemistry—a term actually coined later—display a wealth of new and creative ideas and provide impressive examples of the experimental sixth sense that he developed. His work in this field began very early. On February 13, 1839, a fortnight after he had produced his first photograph, his diary states that he had worked "with great interest and success at the photography and chemical rays." [20] His first paper on the chemical action of the spectrum describes a large number of experiments with inorganic and organic materials. This paper is mainly concerned with the behavior of inorganic substances when exposed to the solar spectrum. Herschel started from the known fact that not all rays of the spectrum have equal chemical effects. The German physicist Johann Wilhelm Ritter (1776–1810) in 1801 and Wollaston in 1802 had found that the violet and ultraviolet end of the spectrum especially affects photosensitive materials. Wollaston therefore called these rays the chemical or actinic rays of the spectrum. As Herschel wrote, Wollaston had shown from experiments with the photosensitive resin of the West Indian guaiacum tree that certain reactions were also caused by the red end. These, however, he had attributed to the strength of the heat rays in this region of the spectrum so that he did not regard them as photochemical reactions in the true sense. Herschel was the first to show that the "negative" end of the spectrum—that is, the region from green to far red—was also chemically active.

To begin with, Herschel measured the intensity and extent of the spectrum on various photographic papers, finding large differences. In some cases, the chemically active portion of the spectrum covered the whole visible region and extended far beyond it at both the red and the violet end. With other sensitized papers, the effect stopped at orange or extended only to the green or blue. The maximum often occurred in the ultraviolet and sometimes in the visible violet or blue. In many cases, several maxima were observed. From these effects he inferred that the degree of chemical effectiveness of any given spectral region was not simply a function of its wavelength or refrangibility but must also depend on other properties of the light rays. The American scientist John William Draper (1811–1882) found that the only light rays causing chemical reactions are those that are absorbed by the photographic medium.

With the aid of a crown-glass prism made by Fraunhofer and a strong collecting lens, Herschel succeeded on July 9, 1839, in producing on silver-chloride paper a picture of the solar spectrum in its natural color. "The result," he writes, "was equally striking and unexpected. An intense photographic impression of the spectrum was rapidly formed, which, when withdrawn and viewed in moderate daylight, was found to be coloured with sombre, but unequivocal tints, imitating those of the spectrum itself." [21] The picture could not be permanently fixed, but Herschel noticed that, if the paper was stored in the dark for a few days, the colors came out more strongly than immediately after the exposure, although they then rapidly disappeared.

Similar experiments had been made in 1810 by the German physicist Thomas Johann Seebeck (1770–1831) and were described by the poet Johann Wolfgang von Goethe in connection with his theory of color.

In Herschel's experiment it turned out that the coloring of the paper did not extend beyond the orange portion of the visible spectrum but gradually faded in this region into a faint brick red. On repeating the experiment, however, Herschel discovered that red light must have acted on the silver chloride as well. Although the paper had taken on a faint brownish background tinge as a result of scattered sunlight that fell on it during the exposure, those portions on which the red light rays had fallen were bleached. Thus red light appeared to have the opposite effect from violet light on the silver layer: whereas the latter blackened the silver, the former whitened it.

By making this observation, Herschel had proved that the red rays of the spectrum are not chemically inactive, as had previously been assumed. He then left the silver-chloride paper in direct sunlight for a short time before exposing it to the spectrum. The exposure gave it a brownish violet color. This time the red rays of the spectrum did not cause bleaching, but a noticeable reddening. When the pre-exposure was intensified by first exposing the paper to violet light or passing the light through an absorbing layer of copper ammonium sulphate solution, the color effect in the red region of the spectrum was considerably enhanced and "a full and fiery red" was produced. By using silver-bromide paper, he finally succeeded in following the chemical effect of the spectrum far beyond visible red light into the infrared.

Herschel also undertook experiments in the ultraviolet region of the spectrum. His earliest tests on the operation of prismatic rays on sensitized paper had shown him that the chemical effect was usually greatest in the violet and extended far beyond it into the ultraviolet. On silver-nitrate paper the chemically active portion of the spectrum was 1.57 times as long as the visual region, on silver chloride 1.8 times, and on silver-bromide, which was exceptionally sensitive, 2.16 times.

Experiments with a high-dispersion prism of flint glass made by Fraunhofer and a lens of crown glass produced under certain experimental conditions a visible light-gray oval spot on the paper beyond the visible violet. Originally separate from the spectrum, this spot rapidly joined up with it and caused intense blackening of the silver salts. Herschel called the new spectral color derived from the ultraviolet region "lavender colour."

Using a special prism-and-lens combination (now preserved at the Science Museum, London), Herschel investigated the simultaneous chemical action of two light rays with different refractive indices on sensitized paper. He wrote: "The results were very striking and beautiful. The blackening power of the more refrangible rays seemed to be suspended over all that portion on which the less refrangible fell, and the shades of green and sombre blue which the latter would have impressed in a white paper, were produced on that portion which, but for their action, would have been merely blackened." [22] Thus there was a kind of superposition whereby the red rays quenched the effects of the violet and blue ones.

Herschel's observation that the latent image can be weakened by a second exposure to red light has come to be known as the "Herschel effect" and is of some importance in the theory and practice of photography. Draper and others discovered that the Herschel effect also occurs in the latent image on a daguerreotype plate. The Herschel effect plays a significant part in the production of direct positives and duplicate negatives.

Herschel tried to use his prism instrument to reveal the Fraunhofer lines as zones of chemical inactivity in the spectrum, but this was first achieved in 1842 by the French physicist Alexandre Edmond Becquerel (1820–1891). Owing to the lack of radiation in the absorption lines, these were expected to appear unblackened in the photographic spectrum—that is, the original color of the photographic paper.

Herschel carried out an extended series of experiments with colored glasses and liquids, in order to find out what chemical effects are caused by light when it is passed through absorbing media after dispersion by a prism. This was a continuation of earlier experiments in which he had measured the absorbing effect of various filters on white (polychromatic) sunlight (see Chapter 2).

He also incorporated in his experiments the invisible heat rays that his father had discovered in 1800 [23] and found that these reached a maximum beyond the visible red, thus confirming a result that William Herschel had obtained by a different method.

Some researches of Herschel's on vegetable substances were described in detail in the second of his two papers in the *Philosophical Transactions* for 1842. A. Hagemann had discovered the sensitivity to light of certain resins as early as 1782 and in this connection had described the properties of guaiacum. As mentioned earlier, Wollaston also experimented with this resin, as did Herschel, who dissolved it in alcohol and covered a piece of paper with the solution, which formed an almost colorless coating. When it was exposed to the spectrum, a fine, blue stripe quickly appeared in the violet and ultraviolet regions. After a longer exposure, dispersed sunlight falling on the paper brought out a brownish-green background tone which spread over the whole surface except for those parts on which the less refrangible yellow and red rays had fallen. When Herschel subjected the guaiacum-coated paper to chlorine vapor immediately before exposure, the spectrum came out in its natural colors as with pre-exposed silver-chloride paper. Only green was lacking, and the yellow was not very marked. When the chlorine treatment was repeated with paper on which the resin solution was still wet, the sensitivity to light was considerably increased. The paper immediately took on a shining blue color which, however, rapidly changed to a brownish- or reddish-yellow in the region of the red rays which now had a strong effect. Thus red light seemed to destroy the blue color, an effect which Herschel found in other vegetable substances as well and which suggested that dyes are decomposed by light of the corresponding complementary color. He found that orange-yellow colors were destroyed by blue light, violet colors by green light, and blue colors by yellow or red light.

As with most sensitive substances, the region of the spectrum having the greatest photochemical effect on guaiacum paper was that of the blue, violet, and ultra-violet (lavender-color, as Herschel called it). By treatment with chlorine, Herschel had extended the effect to less refrangible rays out to the far red. In order to examine the infrared region as well, he now exposed the paper to heat rays after treating it with solutions of guaiacum mixed with soda and of chlorine in water. This produced a greenish or blue background tone which, by heating to 100°C, was changed into a brownish yellow; the effect depended on the precise degree of heating and was absent below a certain threshold temperature.

When Herschel exposed the paper, still wet, to the spectrum, it was bleached in the region of yellow and red light, as in the case of the silver chloride experiment. At the same time, the heat radiation in the infrared affected the paper and caused evaporation of the solution at certain places where the effect was a maximum. A discontinuous thermal spectrum was produced, but this disappeared when the paper was dried, whereas the chemical effect of the visual rays persisted.

As the example of guaiacum shows, Herschel carried

out a photochemical exploration of the various regions of the spectrum. He called his pictures of the heat spectrum "thermographs." They were produced by a bleaching process, whereas the chemically active rays in the violet and ultraviolet caused the guaiacum paper to undergo a change in color. In the course of these thermographic experiments, Herschel made the very important discovery that the heat rays apparently were not confined to the zone beyond the far red but overlapped with the other rays far into the visible region.

Finally, Herschel experimented with floral dyes: "The petals of fresh flowers . . . were crushed to a pulp in a marble mortar, either alone or with addition of alcohol, and the juice expressed by squeezing the pulp into a clean linen or cotton cloth. It was then spread on paper with a flat brush, and dried in the air . . . or at most with the gentle warmth which rises in the ascending current of air from an Arnott stove." [24]

The photographic papers produced in this manner had to be used as rapidly as possible, because the dye was decomposed by the action of air and moisture. The addition of alcohol, acids, or alkalis frequently led to a change in color or complete discoloration; often the original color returned after the paper had been dried, but in other cases the plant extracts took on entirely different colors on the paper and kept them. Herschel found, for example, that the dye extracted from the dark-red damask rose changed into blue, as did the extract from red poppies, whereas the juice of a fine pink tulip produced a dirty bluish green.

When papers treated with the juices of plants were exposed to direct sunlight, the colors were usually bleached as they were in the earlier experiments with heat rays, although the latter now had absolutely no effect. The exposure times varied considerably. In extracts from the yellow petals of the shrub *Corchorus Japonica,* for example, an exposure of ten minutes was enough to cause a marked bleaching of the yellow coating. After half an hour, the paper was totally discolored. With other plants— the Oriental poppy (*Papaver Orientale*), for instance—a week's exposure was sometimes needed and even then one often could not tell which parts of the paper had been acted on by light until the paper had been treated with weak acids.

Total discoloration of the paper did not always occur. Frequently the exposed places retained a "residual tint," as Herschel called it, which was insensitive to the further action of light. The most rapid bleaching appeared to occur with those dyes whose colors were complementary to those of the incident radiation, an effect that had also appeared when guaiacum was used.

Using prismatic light, Herschel demonstrated that floral dyes, unlike silver halides, usually react chemically only with rays from the visible spectrum. The spectrum recorded on *Corchorus Japonica* paper showed several intensity maxima. The green, blue, and violet regions were bleached the most, whereas the red zone showed hardly any effect. With paper soaked in the juice of a red stock (*Matthiola annua*), the effect was concentrated in the yellow and red spectral regions and disappeared completely toward the blue, reaching a second but much lower maximum in the violet.

An interesting further proof of the fact that the spectrum can be followed beyond the violet end optically as well as chemically was provided by Herschel's experiments with an alcoholic tincture of turmeric root (*Curcuma longa*). When a piece of paper coated with the tincture was exposed to the spectrum, a faint but sharply defined pale yellow stripe became visible, stretching beyond the violet into the "lavender-coloured" region. A similar, though weaker effect was obtained with paper coated with the dye of a purple-red dahlia. "And if such," Herschel wrote, "rather than lavender or dove-colour, should be the true colorific character of these rays, we might almost be led to believe (from the evident reappearance of redness mingled with blue in the violet rays) in a repetition of the primary tints in their order, beyond the Newtonian spectrum . . ." [25]

Herschel made experiments with a great variety of vegetable dyes. These, for all their diversity, represent but a tiny part of the great wealth of possibilities that had been opened up in this field. Mary Somerville and Robert Hunt made similar investigations, but Herschel undoubtedly deserves the credit for having created a broad foundation for further work on the chemical effects of light.[26] His work produced a singularly striking demonstration of the need for a close amalgamation of physics and chemistry at a time when the connection between the two was by no means as obvious as it is now.

Herschel spent only five years, from 1839 to 1844, in intensive work on photography and photochemistry, and this period included a number of other tasks, such as official duties that could not be postponed. He had come upon photography more or less by chance, without realizing what the scope of his researches was to be or their importance in his scientific life. His enthusiasm for the effects of light on sensitized papers led him on from one experiment to another, and finally to insights and discoveries which have proved fruitful in chemistry and physics as well as in photography.

7

The Collingwood Period

✦
✦
✦

On January 1, 1840, a strange ceremony took place in the garden of the Herschel house at Slough. On the lawn lay the 40-foot-long tube of the great reflector that William Herschel had first pointed at the heavens half a century earlier. The instrument had been out of use for over thirty years, and the enormous wooden framework from which the tube had been suspended by strong ropes had suffered so much from its long exposure to the weather that John Herschel had to have it dismantled in the autumn of 1839. Only the iron tube had held out against decay. Herschel desired it to be preserved as a monument to remind later generations how the depths of the universe had been penetrated from this tiny piece of garden.

At noon on New Year's Day John Herschel assembled his family inside the disused tube. He read a poem that he had written as an obituary for the time-honored instrument. The grinding and polishing tools that William Herschel had used in the construction of the telescope were placed inside the tube, and it was then ceremoniously closed and sealed.

This sentimental little ceremony is touching evidence of the almost childlike love and respect with which the son in middle age still regarded his famous father. It also marked John Herschel's formal farewell to his own observing activities. "I have no telescope in use now," he had written to a friend shortly after his return from the Cape.[1] The 20-foot reflector which had been his main instrument in South Africa lay unused in the cellar at Slough, never to be set up again. The other telescopes that William Herschel had used had also been dismantled and were merely museum pieces adorning the house and garden at Slough. Only one 7-foot telescope, the one used in the discovery of Uranus, was later set up again and used for occasional observations.

But John Herschel, although he now considered his activities as an observing astronomer ended, could not bring himself to give up observation altogether. In the spring of 1840 he surprised the astronomical world with a discovery that had eluded all previous observers, even though it concerned one of the brightest objects in the sky. He found that the bright star Betelgeuse (Alpha Orionis) displayed periodic variations in light—that is, it is a long-period variable star.[2]

Herschel seems at first to have intended to continue in England the magnitude estimations that he had begun in South Africa. The work could have even been done without a telescope, and the results would have been most valuable in view of the lack of knowledge of this field and the current neglect "of this highly interesting branch of Physical Astronomy," as Herschel calls it in his paper on Alpha Orionis. However, he made only the one investigation. Many activities engaged his time in the first years after his return to England, and his uncertain state of health made long night watches in the open air inadvisable. John Herschel had never had a very strong constitution and lacked the robust good health which his father enjoyed up to an advanced age. He was susceptible to the effects of chills, easily caught on damp or frosty observing nights. At Feldhausen, despite the warm climate, he had contracted rheumatic troubles and a tendency toward severe attacks of bronchitis, both of which plagued him especially in later life. In January 1839 he wrote to Schumacher: "I fear that my health will no longer suffer me to indulge the hope of prosecuting these enquiries myself further in this hemisphere. To my no small annoyance I find that night exposure at least in the winter season is more than I can now face, having been of late a sufferer from severe rheumatic affections which warn me pretty forcibly to desist." [3]

The spring of 1840 brought a change in the life of the family. Herschel had decided to give up his father's house at Slough and to escape from an area which was becoming increasingly urbanized as a result of the rapid growth of London and the construction of a railway line. His health was probably one cause of his aversion to city life, but the move seems to have been motivated by two other considerations, one scientific and the other connected with his family. He wanted a refuge in which he would at last be able to work up the results of his South African expedition undisturbed. His letters to various friends show how much he longed to begin this work. He wrote to the Scottish scientist James David Forbes (1809–1868) in March 1839: "I have been entirely disabled from advancing a step in the reduction of my Cape observations, which remain precisely (with one most trifling exception) in the state in which they were when I left the Cape." [4] The family consideration was the fact that the old house at Slough was far too small for a family of seven children. A fourth daughter, Maria Sophia, was born after the return from South Africa.

Herschel began looking for his new home in the summer of 1839 and found a fine and spacious property near the village of Hawkhurst in Kent. The house, situated in a large park surrounded by hilly countryside, was an ideal retreat for a secluded life of research. Its size and comfort may have reminded Herschel of Feldhausen. The contract of purchase was soon completed, and on April 3, 1840, the family moved into the new home, which was called Collingwood, the name of its previous owner.

He started work on his report of the results of the Cape expedition without delay. An abundance of observational material, the result of 400 nights' observing at Feldhausen, had to be evaluated. The work involved, to begin with, lengthy and largely purely mechanical computations for reduction, a tedious, time-consuming task to which he devoted himself with immense patience. His friends, especially Whewell, complained vigorously about the fact that he was devoting his time and energy to this work, which could have been carried out by a competent assistant. But Herschel wanted to check the correctness of the minutest calculations himself and would not hand over the observing logs to anybody.

Slowly but steadily the book progressed. Various diversions continually threatened the work, since his researches in terrestrial magnetism and photography were partly carried out during this period, but he steadily continued computing and writing, often working far into the night. This kind of life had a very damaging effect on his health, and he was often utterly exhausted. In 1844 he wrote to Colonel Sabine: "I feel my health rapidly breaking and I have many and distinct warnings that what I have to do I must do quickly, that *time* is the stuff of which life is made." [5] Although Herschel often gave expression to his momentary feelings in a dramatic and impulsive manner, this statement does indicate something of the effort required for the book that was to be the crowning glory of his astronomical enterprise.

Finally the work was done. With great relief Herschel put down his pen on March 7, 1847, his fifty-fifth birthday. In joyful mood he noted in his diary: "A pleasing and happy day. Papa wrote in solemn style the word Finis to 'his book' and Mama and the little ones crowned him with bays . . ." [6]

A generous subsidy from the Duke of Northumberland, a patron of science, enabled Herschel to produce the book in an attractive and worthy form. The large quarto volume was published in London in 1847 with the comprehensive title *Results of Astronomical Observations Made during the Years 1834, 5, 6, 7, 8 at the Cape of Good Hope; Being a Completion of a Telescopic Survey of the Whole Surface of the Visible Heavens, Commenced in 1825.*

Although the observations recorded were described in Chapter 4, the book itself is worth summarizing here.

In an extensive introduction, Herschel explains the motivation and purpose of his enterprise and discusses the equipment of the expedition. He describes the voyage to Cape Town and the erection of the observatory at Feldhausen, with details about the instruments and the polishing and storage of the mirrors, and gives an account of the climate.

The nebulae and star clusters which Herschel discovered and observed in his sweeps of the southern sky with the 20-foot telescope are the subject of the first chapter. The most conspicuous objects are briefly described, and a number of drawings illustrating this and later chapters appear at the end of the book. The Great Nebula in Orion is discussed at length and illustrated by a remarkably detailed drawing. Another section, illustrated with a chart, is devoted to the extended nebulosity surrounding Eta Argus (Eta Carinae); it includes a catalogue of 1200 stars within the nebulosity. This is followed by a general catalogue of 1707 nebulae and clusters in the whole of the southern hemisphere, with the objects placed in order of right ascension and reduced to the equinox of 1830. Finally there is a statistical discussion of the distribution of star clusters and nebulae and a classification of nebulosities according to their visible features. The chapter concludes with a description of the two Magellanic Clouds and a list of the stars, clusters, and nebulae within them.

The list for the Small Cloud contains 244 entries, that for the Large Cloud 919. The section on the Magellanic Clouds is illustrated with two impressive charts.

The second chapter of the *Results* deals with double stars. Observations with the large telescope had produced a catalogue of 2102 binary systems, arranged like the nebulae in order of right ascension and reduced to the equinox of 1830, which takes up most of the chapter and is followed by a list of 1081 double stars for which Herschel had measured the separation and position angle with the 7-foot equatorial. This valuable list is completed with notes on the observing conditions and other details of the measurements. The observational data secured with both telescopes are combined into average values in a list of 417 double stars. Further details on a number of especially noteworthy objects conclude the chapter. "Of Astrometry," the third chapter, discusses a topic to which William Herschel had already given special attention. The term "astrometry" (measurement of stars) is somewhat misleading, since Herschel used it with a different meaning from that given it today. In modern usage the term refers to the study of the real and apparent motions of celestial objects and the determination of the positions of stars on the celestial sphere; Herschel's investigations were concerned with the determination of the apparent brightness of the stars, a subject that is now called "photometry" (measurement of light). In the introduction to this chapter Herschel describes his father's investigations and his own methods, which have already been outlined (see Chapter 4). He gives a list of 2341 estimations of brightness carried out in 46 sets of observations (sequences), in which the stars of each sequence are arranged in order of apparent brightness; for each star a numerical value of the brightness, expressed in magnitudes, is given, based on estimations in the sequences in which it occurs. Since there are many stars that occur in several sequences, the magnitudes given for one and the same object often differ. To obtain the real magnitude of a star, independent of the sequence in which it occurs, Herschel collected all the stars occurring in several sequences into another list and took for each one an average of the values found in different sequences. This then gave a mean estimate of the apparent magnitude relative to the whole set of stars observed. Another part of this photometric chapter treats of Herschel's experiments with his astrometer. The results are embodied in a table containing some 70 of the brighter stars in the southern sky.

The fourth chapter begins with a description of Herschel's method of star gauges, which was the same that his father had used for the northern hemisphere. All the stars visible in the field of view of the 20-foot telescope were counted and the numbers listed in a series of zones of right ascension separated by successive intervals of 10 minutes of arc. A detailed description of the general characteristics of the southern portion of the Milky Way follows.

Halley's comet is the subject of the fifth chapter. It includes Herschel's observations of the comet, which had been visible from Feldhausen from the end of October 1835 to the beginning of May 1836, with two months' interruption at the time of perihelion passage, as well as a general discussion of the nature of the comet. There are a number of drawings illustrating this chapter.

The last two chapters describe investigations that did not belong to the main body of Herschel's research program. One summarizes the results of observations of the seven known satellites of Saturn which Herschel carried out between 1835 and 1837. The other describes a series of observations of sunspots and attempts to develop a theory of them.

Herschel's *Results* aroused immense admiration and enthusiastic recognition. The Royal Society marked the achievement by again awarding Herschel the Copley Medal in 1847.

One of the first copies of the book was sent to Caroline Herschel at Hanover. In an accompanying letter, Herschel wrote: "You will . . . have in your hands the completion of my father's work—'The Survey of the Nebulous Heavens.'" The indomitable old lady was then physically feeble but mentally as fit as ever and highly interested in the doings of her relatives. She died in Hanover on January 9, 1848, in her ninety-eighth year.

During the years that he devoted to the preparation of his *Results*, Herschel never lost sight of day-to-day events in astronomy. He had set up the 7-foot telescope from Slough on a small observing platform on the roof of his house at Collingwood and used it to watch for special celestial events and to make regular observations of the annual recurrences of the Perseid and Leonid meteor streams and of any faintly visible comets. The great comet of 1843, the biggest of the century, appeared on March 17 as a ". . . remarkable streak of light under Orion"[7] and was followed for several evenings. There are further diary entries on cometary observations between 1844 and 1846, and Biela's comet was repeatedly observed from the end of January to February 24, 1846.

Scientific visitors were frequent at Collingwood. Among them was Encke, who was Herschel's guest in September 1840. On July 12, 1842, Bessel came to spend two days in this hospitable household on his return journey from the British Association's annual meeting at Manchester. During his visit, Bessel expressed the belief that there must be another planet beyond the orbit of Uranus. Certain perturbations of the latter's orbit had been found which could not be satisfactorily explained as a result of the action of the other known planets of the solar system, but only by taking into account the gravitational effect of a still unknown "trans-Uranian" planet.

Herschel greeted Bessel's idea with enthusiasm. In August he wrote to Baily of the need to organize a full-scale hunt for this still hypothetical celestial body. The ecliptic was to be divided up into zones, and astronomers were to examine carefully all the stars down to the eleventh magnitude, each in a particular zone.

The search was not carried out, but several years later, and in a different way, the chase was taken up by two astronomers, Urbain Jean Joseph Leverrier (1811–1877) in France and John Couch Adams (1819–1892) in England. The work of these two men led to one of the great triumphs of computational astronomy. Both derived the orbital elements of the unknown planet by purely mathematical methods and predicted its position at a given time.

In 1845, Adams, who was still a student at Cambridge, sent his results to the Cambridge Observatory, and James Challis (1803–1882), the director, passed the paper on to Airy, then Astronomer Royal. Unfortunately, both these men were skeptical about the young and unknown Adams and did not take up the search for the new planet immediately. So much precious time was lost in cross-questioning and detailed inquiries into Adams' methods of computation that the occasion for a telescopic search with good prospects of success was missed because the planet had by then moved into the daylight sky.

Adams had begun his computations in 1843. Two years later, Leverrier, without having heard anything of Adams' work, also set about computing the orbital elements. In the summer of 1846, he sent his results to the German astronomer Johann Gottfried Galle (1812–1910) at the Berlin Observatory. On September 23, Galle noticed in the immediate neighborhood of the predicted position a faint object of the eighth magnitude which did not occur on the star charts of the Berlin Academy. On the following night, its position among the fixed stars had slightly altered, proving that the object was indeed the new planet, afterward named Neptune. In effect, it was discovered at the astronomers' writing desks, because its observation in the telescope merely confirmed what had been predicted by computation.

When Challis and Airy heard of Leverrier's computations, they were impressed by the remarkable agreement with those of Adams, and immediately undertook the search for the new planet. They did not find it, although Challis had actually noted Neptune in his observing book as a fixed star on two occasions—August 4 and 12, 1846—before Galle's observation.

There now arose a lively and in some quarters a passionate debate over the priority in the discovery of Neptune. Herschel took a great interest in the whole affair, which threatened to become a topic of international controversy. In his speech on handing over the office of president of the British Association to the geologist Sir Roderick Impey Murchison (1792–1871) at the summer meeting of 1846, he referred to the still undiscovered planet: "We see it as Columbus saw America from the shores of Spain. Its movements have been felt, trembling along the far-reaching line of our analysis, with a certainty hardly inferior to that of ocular demonstration." [8]

Herschel had come very close to discovering Neptune himself, in the same way that his father discovered Uranus, long before anyone had thought of the existence of another member of the planetary system. In a letter to Wilhelm Struve in December 1846 he wrote that, on the night of July 14, 1830, he had swept a zone of the sky only half a degree north of the place where, according to the computations, the planet must then have been. The magnifying power of his telescope would have been sufficient to show Neptune as a small but recognizable disk and thus to reveal that it was a planet. "But it is better as it is. I should be sorry it had been detected by any accident or merely by its aspect. As it is, it is a noble triumph for science." [9]

Herschel used all his efforts to put forward a fair and balanced point of view in the heated controversy over priority for this important discovery. Unhampered by national prejudice, he recognized Leverrier's great achievement and agreed that he deserved the credit on the grounds that his work had been the first to be published and the discovery had actually been made on the basis of his results. On the other hand, Herschel pointed out that Adams had reached the same result quite independently of Leverrier and in fact earlier, so that, as far as the purely scientific aspect of their discovery was concerned, both were equally deserving of recognition for what they had done. He recommended to the Council of the Royal Society, on which he was serving, that the Copley Medal be awarded to Leverrier, but he also recommended to the Royal Astronomical Society that Adams receive its Gold Medal. In the end, the latter society avoided taking sides by the stratagem of replacing the Gold Medal by awards of special testimonials to a number of astronomers, including both Adams and Leverrier, and also Herschel, for his *Results.*

Leverrier and Adams themselves stayed aloof from the controversy and always treated each other with courteous regard—a circumstance which may be due in large measure to a personal meeting of the two which took place at Collingwood. Herschel had arranged this meeting in order to wipe out any hard feeling between the competitors by means of friendly human contact in a private house and, as it were, on neutral ground. Herschel's simple sense of justice and his friendly warm-hearted ways did even more than the prestige of his scientific authority to make the meeting an unforgettable experience for both Leverrier and Adams. The encounter took place on July 10, 1847, on the occasion of the Oxford meeting of the British Association. Wilhelm Struve and Charles Pritchard (1808–1893), the Savilian Professor of Astronomy at Oxford and a good friend of Herschel's, were also present.

Much as Herschel desired peace and seclusion for the prosecution of his work, any visitor who cared to journey from London to the astronomer's house received a warm welcome. Airy, an especially faithful and constant friend of Herschel's, frequently came from Greenwich with his family to spend several days at Collingwood. Babbage, Whewell and Peacock, Herschel's old Cambridge friends, were other welcome guests, as was Pritchard, who, in addition to his professorship at Oxford, was headmaster of the Grammar School, Clapham, an excellent private school which all three of Herschel's sons attended. Another visitor to Collingwood was the Scottish writer Mrs. Mary Somerville, who had made a name for herself with her books and articles on physics and astronomy. Her most important work was an English translation of Laplace's *Mécanique Céleste*, which had appeared in 1831 under the title of *Mechanism of the Heavens* and had been reviewed at length by Herschel in the *Quarterly Review*.

This was not Herschel's only book review; three others appeared in the *Quarterly Review* and also in the *Edinburgh Quarterly Review*: one was of Humboldt's *Kosmos*, the second discussed Whewell's *History and Philosophy of the Inductive Sciences*, and the third concerned a book on the theory of probability by the Belgian statistician and astronomer Lambert Adolphe Jacques Quételet (1796–1874). Other writings by Herschel included obituaries of Baily (1844) and Bessel (1846)[10] and a major article, "Physical Astronomy," which appeared in the *Encyclopaedia Metropolitana* in 1845.[11]

These articles, which were partly occasioned by Herschel's official duties as president of the Royal Astronomical Society, were more or less peripheral to his main literary activities during this period. Important among the latter was the new and enlarged edition of the *Treatise on Astronomy*. Although the new edition, called *Outlines of Astronomy*, was still intended for the educated layman, it cannot be regarded as a popular account as that term is used today. To read this tome, some 700 pages in length, requires a high degree of concentration and a willingness to go through the abundance of material with careful attention. This applies especially to the long chapter on the perturbations of planetary orbits and their causes. Of this chapter, Herschel wrote in the preface: "The chief novelty in the volume, as it now stands, will be found in the manner in which the subject of perturbations is treated. It is not—it cannot be made *elementary* in the sense in which that word is understood in these days of light reading. The chapters devoted to it must, therefore, be considered as addressed to a class of readers in possession of somewhat more mathematical knowledge than those who will find the rest of the work readily and easily accessible . . ." [12]

The book was enormously successful. It went through twelve editions between 1849 and 1873, and an edition was published in New York as late as 1902. It was translated into many languages, including Arabic and Chinese. This brilliant success may have been partially due to the prestige of Herschel's name, but its chief cause was probably the fact that no equally comprehensive general account of astronomy was available in the mid-nineteenth century, either in England or elsewhere. The sort of collective textbooks with contributions from various authors that nowadays provide a comprehensive survey of nearly every field in the arts and sciences were almost unknown in Herschel's day. Furthermore, specialization had already developed to such a degree, especially in science, that it was becoming increasingly difficult for a single spokesman to survey the whole of his field, let alone deal with it adequately. In his *Outlines*, Herschel succeeded in producing such a comprehensive survey, which did much more than merely summarize the results; it was described by Agnes M. Clerke as ". . . perhaps the most completely satisfactory general exposition of a science ever penned." [13] The only book of its time that can be compared with it is Humboldt's *Kosmos*.

Despite the difficulties of its style, this book, like most of Herschel's publications, exerts great charm over the reader, because every line expresses the author's enthusiasm for his subject. He says, for example, of the famous southern star cluster Kappa Crucis: "Eight of the more conspicuous [stars] are coloured with various shades of red, green and blue, so as to give to the whole the appearance of a rich piece of jewellery." [14] And on the observation of certain star clusters and nebulae in Lord Rosse's reflector he writes: "The sublimity of the spectacle afforded by that instrument of some of the larger globular and other clusters enumerated in the list given in Art. 867 is declared by all who have witnessed it to be such as no words can express." [15]

Outlines of Astronomy and *Results* were not Herschel's only writings during the years following his move to Collingwood. In January 1845 he wrote to the English botanist John Lindley (1799–1865) that he had been working for eighteen months on a book on meteorology.[16] This was intended as a popular introduction to the subject and was to provide an insight into its foundations in the form of an imaginary correspondence between a father and his son who was on a sea voyage. The plan was dropped for the time being because of more pressing commitments, especially the preparation of the *Results*, but the project was not forgotten.

Herschel's meteorological researches at Feldhausen and his part in establishing regular and systematic weather observations at stations throughout the British possessions have been discussed earlier. In 1847, he was asked by the First Lord of the Admiralty, Lord Auckland, to act as editor for a proposed "Manual of Scientific Inquiry" and to contribute an article on meteorology to it. The manual was intended to be a textbook for cadets at the Royal Naval College, Greenwich, and was to provide basic information on the scientific subjects that concerned naval officers. Articles on various aspects of astronomy, physics, and mathematics were contributed by Airy, Whewell, the geologist Adam Sedgwick (1785–1873), Sabine, and Beaufort. A *Manual of Scientific Enquiry* was published in 1849,[17] the same year as the *Outlines*. Herschel's article on "Meteorology" was reprinted in the eighth edition of the *Encyclopaedia Britannica* [18] and later published as a separate volume.[19] A shortened and popularized version of it, entitled "On Weather and Weather Prophets," was included in Herschel's *Familiar Lectures on Scientific Subjects.*[20]

Herschel wrote two further articles for the *Encyclopaedia Britannica*, a long one on "Physical Geography" [21] and a shorter one "Telescope"; [22] both were by-products of *Outlines*.

Another publication of Herschel's during this period deserves to be described more fully because it helped to prepare the ground for a fundamental discovery made in 1852 by the physicist and mathematician George Gabriel Stokes (1819–1903), which led to the development of a new field in physical optics: that of fluorescence. Herschel's contribution to this subject shows that he never confined himself to scholarly surveys for encyclopedias but continued to engage in wide-ranging experimental research. His work in photography and photochemistry shows what a passionate experimenter and research worker he was.

The *Philosophical Transactions* for 1845 contains two short papers, bearing the somewhat peculiar collective title of *Amorphota* (Greek ἀμόρφωτα, various things), which summarize the optical properties of certain chemical solutions.[23] The first paper describes experiments for which Herschel prepared a watery solution of quinine sulphate and tartar and poured it into a glass cylinder. When the cylinder was held up against the daylight or a bright surface, the liquid either appeared completely colorless or suddenly became a luminous sky blue, according to the angle at which the light rays entering the cylinder were viewed. When the solution was viewed from above, so that the line of sight ran nearly parallel to the inner wall of the cylinder, the coloration appeared most strongly in those regions of the surface of contact between the liquid and the wall where the light was directly incident. When it was viewed at a more oblique angle, the blue zone spread out into a paler band.

To prove that this was not simply a reflection effect caused by the passage of light from a denser medium (glass) into a rarer medium (the solution), Herschel allowed sunlight to fall on the liquid while it was being poured into another vessel and observed the same luminous blue glow from the streaming liquid particles. The effect was most impressively demonstrated when the liquid was poured into another, previously wetted cylinder in such a way that it trickled down the walls in the form of a thin film. This gave the impression of a body of liquid that was colored throughout, although actually the color was only on the surface (Herschel calls it a "superficial colour" in his article).

Examination with a prism revealed that red light was hardly at all involved in the effect, whereas blue and violet light were very much involved. No signs of polarization could be detected. Herschel carried out experiments with other substances and found that the effect occurred only in acid solutions.

The second paper describes further experiments with solutions of quinine sulphate, by means of which the light colored by the solution was analyzed. Herschel called this type of dispersion of light "epipolic dispersion" (Greek ἐπιπολὴ, surface), since the effect was observed only at the surface of the liquid. Later this expression was replaced by the term "fluorescence" suggested by Stokes.

Herschel does not seem to have carried out many more experiments in this field. In a postscript to another paper there is a brief mention of esculin, an extract from the horse chestnut, which also shows fluorescent effects. Similar experiments had been made on chlorophyll many years earlier by the physicist David Brewster. How Herschel got the idea for his experiments is not known; presumably it developed from his researches on the photochemical properties of organic substances, already described, which were still going on at this period. In view of his experiments with vegetable substances, it is strange that he overlooked the fluorescent properties of chlorophyll.

The experiments of Brewster and Herschel were repeated and extended by Stokes. Though Stokes does not really rank as the discoverer of epipolic dispersion, or fluorescence, he was the first person to investigate the phenomenon thoroughly and to understand its nature clearly. In a paper communicated to the Royal Society in 1852,[24] he showed that fluorescence is due to intrinsic emission of light by the medium after it has been excited by the incident radiation. He explained the emission by supposing that when the molecules of a fluorescent substance are set into vibration by light they immediately re-emit the light rays that they have absorbed. From experiment he discovered the rule named after him, according to which the emitted light has a longer wavelength than, and thus a different color from, the absorbed light.

Herschel's work on the fluorescence of quinine sulphate was his last major paper in the field of physical optics. The duties of his various offices were occupying him to an increasing extent. In 1847 he was elected president of the Royal Astronomical Society for the third time. The situation that led to this was similar in many respects to that which had faced the society when it offered the presidency to William Herschel in 1821: it was at a crisis in its development and needed the prestige of an illustrious name to assure its continued existence. The English mathematician Augustus De Morgan (1806–1871) expressed the situation in a letter to Captain Smyth in which he wrote: "If we give bail that we won't let him do anything if he would, we shall be able to have him, I hope. We must all give what is most wanted, and his name is even more wanted than his services. We can do without his services, not without loss, but without difficulty. I see we shall not, without great difficulty, dispense with his name." [25]

Herschel felt that the responsibility was incumbent on him as one of the founders of the society. When he handed over the office to his successor two years later, he had the satisfaction of knowing that the crisis had been overcome. However, he firmly declined an offer of the presidency of the Royal Society made to him in 1848, just as he had the first such invitation.

In the midst of his busy scientific career during these years Herschel enjoyed an undisturbed harmony at Collingwood which presents a delightful contrast to his other existence. Herschel always had time for his own family, even during his years of intensive scientific activity. Every visitor who was privileged to enjoy the hospitable atmosphere of Collingwood was made to feel at once as though he were one of the family and was kindly and unobtrusively absorbed into the peaceful, friendly routine of the astronomer's household.

Sir William Rowan Hamilton once summarized in verse his impressions of a visit to his friend's house:

> Where all things graceful in succession come;
> Bright blossoms growing on a lofty stalk,
> Music and fairy lore in Herschel's home.[26]

Herschel's love of the arts has already been mentioned. Laconic as his diary is, he often noted in it visits to concerts and operas in London. He also frequented the house of Freiherr von Bunsen, a great connoisseur and patron of the arts, who used to arrange small but very select recitals in his house. There, on one occasion, Herschel made the acquaintance of the German composer Felix Mendelssohn and admired his virtuosity as a pianist. The Herschel daughters, almost all of whom were musical, were given lessons by the best teachers in London, and many evenings at Collingwood were spent in communal music-making. Herschel himself played the violin and flute. Much time was also devoted to drawing and painting lessons, and his eldest daughter Caroline developed into an unusually talented artist. A portrait of her father which she painted after his death, partly from memory and partly from an 1864 photograph, is one of the best likenesses of Herschel in existence.[27]

The place that music held in the education of Herschel's daughters had nothing to do with the conventional ideas of female upbringing of a hundred years ago but corresponded closely to the girls' natural talents. His three sons received at Pritchard's boarding school at Clapham a thorough education in which classical and scientific studies were judiciously combined. Herschel always made it his concern to watch over their scholastic progress. His efforts to teach them ancient languages ranged from light-hearted translations of English nursery rhymes into Latin to communal readings of the Greek and Roman classics. An important role was naturally played by mathematical problems and by experiments in physics and chemistry, demonstrated with enormous skill by Herschel and usually witnessed by the whole family.

His few hours of leisure Herschel devoted primarily to poetry—his own and translations. In 1842 he wrote a verse translation of Friedrich von Schiller's "The Walk," [28] a poem he particularly liked because its evocations of nature reminded him of his own walks in the delightful countryside around Feldhausen, which he often recalled with a certain degree of nostalgia. He also translated other poems by Schiller, among them "Funeral Dirge of a Nadowessie," "Saying of Confucius," "Dithyramb," as well as Gottfried August Bürger's "Leonora," and Dante's *Inferno*. Some of these translations were published in the magazine *Good Words*. Some also appear in the volume of *Essays* published in 1857.[29] His own poems are not of great literary value, but they provide a revealing glimpse of his ideas and feelings. Though some passages are overladen with sentimentality, others charm the reader by the sublimity of their ideas and their simple but impressive language. The final verse of the elegy "On Burning a Parcel of Old Manuscripts" runs:

Enough if cleansed at last from earthly stain,
My homeward march be firm, and pure my evening sky.[30]

Herschel's religious faith, which is echoed in these words, was very marked. He became a devout Christian, although it was not until he reached manhood that he accepted the Christian faith without reservation. His home background had contributed little religious training. His mother belonged to the Church of England, but she seems to have exercised little religious influence over her son. His father came from a family that had been loyal to the Protestant faith ever since the Reformation, but William Herschel's church ties, if they ever existed, seem to have loosened at the time he moved to England. His post of organist at Bath was regarded purely as a means of earning a living; his choice of it does not seem to have been swayed significantly by any religious consideration. In any case, he never expressed himself on the matter. His desire that his son become a clergyman, mentioned in Chapter 1, was also not evidence of Christian beliefs, since it was based on strictly practical considerations.

Nevertheless, William Herschel was by no means irreligious. An admirer of the philosopher John Locke, he was strongly influenced by Locke's ideas on deism. He believed in the working of a divine providence whose will led to the existence and operation of an all-embracing, rational, universal order. But his religion was wholly contemplative and expressed itself purely in an admiring study of this universal order as he saw it displayed in the tiniest crystal and the remotest stellar system.

With this background, it was natural that John Herschel's own early religious ideas were also along partly pantheistic and partly deistic lines. The decisive turn did not come until his marriage. The daughter of a Scottish Presbyterian minister, his wife had grown up in an environment strictly oriented toward the Church and had absorbed Christian belief as an integral part of her personality. Though she never made any attempt to force her convictions on her husband, it was none the less because of the quiet influence of her personal beliefs that Herschel soon committed himself to a clear acknowledgment of Christian faith. A man of delicate feeling and tremendous sensitivity of soul, he may have felt an emptiness and isolation in deism. This rationalistic creed, based on the philosophy of the Enlightenment, rejected biblical revelation and led away from the sources of the Christian mode of life to a purely intellectual humanism. Herschel was by no means an adherent of the general cult of progress which had such a strong influence on the science of his time. He was conscious of the imperfection of human efforts and of the doubtful character of human successes and triumphs, although his own life had been so rich in outward successes and triumphs.

His religion, however, did not make him passive or resigned; on the contrary, he led a full and active life.

He enjoyed the good fortune of a cheerful, untroubled family life with its various pleasures both great and small, and he was distressed when compelling duties kept him away from the family circle even for a few days.

When the family moved to Collingwood, it consisted of three sons and four daughters. Four more daughters were born during the next six years. On February 21, 1841, Herschel wrote in his diary: "This day at 4 p.m. Margaret gave birth to our 5th daughter and 8th child (to be named Amelia)." [31] Julia was born on September 16, 1842; Mathilda Rose on September 4, 1844; and Francisca during July 1846.

The last child, Constance Anne, was born on May 29, 1855. She is of especial interest in her capacity as chronicler of the Herschel family: in her book, *The Herschel Chronicle*, published in her old age, she preserved a unique series of documents relating to William and Caroline Herschel. [32] She seems to have planned a similar book about her father but unfortunately the plan was not carried out. She left only a fragmentary sketch of John Herschel's life, which was privately printed at Cambridge in 1938. Incomplete as it is, this little book, which is quite rare, has a certain charm because it comes from an authentic source and gives the reader an insight into Herschel's personal and domestic life that would be hard to achieve in any other way. In it she says:

"Everything possible was done for the occupations and amusements of the children. . . . The billiard room was divided by a partition; the portion nearest to the study being fitted as a laboratory to which the boys had free access, and the outer part opening on the garden was arranged for carpentry. Here Sir William Herschel's own lathe, on which he and his brother used to turn the eyepieces for the telescopes, was put up and was in frequent use.

"For the girls, besides the regular schoolrooms, a room

was set apart for a studio with a northern aspect and a high window. And for their own amusement they contrived a clay-modelling department under the terrace. Though not much attention was paid to sport or games, yet bars and ropes for gymnastics were put up in a big barn, and there was splendid skating on the large pond in winter. In later years croquet and archery came into favour and the girls rode a good deal. . . .

"There was one ceremony never omitted at Collingwood. It took place on the New Year's Eve. Everyone assembled in the hall at midnight to listen to the church bells ringing in the New Year and then, after exchanging hearty greetings, passed on to the drawing room to sign the Christmas card. These cards, upon which the events of the year are depicted amongst various allegorical devices, were designed by Louisa up to the time of her marriage and carried on afterwards by her sisters more or less successfully. . . ." [33]

The interplay between the pleasures of such a family life and the earnest, silent scholar's life dedicated to work developed in John Herschel a personality that enchanted those of his contemporaries who knew him well. When, shortly after Laplace's death, Charles Pritchard asked the famous French physicist Biot who he considered the most worthy successor to Laplace, Biot replied: "If I did not love him so much, I should unhesitatingly say, John Herschel!" [34]

8

Public Office and Last Years

✦
✦
✦

In the last quarter of a life that had been entirely devoted to science, Herschel, at the age of fifty-seven, made a decision that seems strange in the light of all that preceded it. Though he had always made every effort to avoid public offices and similar commitments, he suddenly conceived the idea of taking up such an office; furthermore, the office was far removed from his scientific interests. The background is partly explained in a letter to his daughter Caroline written in January 1855, several years after his appointment.

"Some years ago when a little wee-blue-diabled about worldly matters I talked to my good old and now lost friend Jones * about them. I asked him if he thought there were any position in the Public Service he thought I could apply for. His answer was—rough and strong and to the point: 'I had rather see you in your grave!!' Some time after he added: 'The only thing fit for you is the Mastership of the Mint. But that is a political office.' Judge of my amazement and how I was taken aback when one fine morning—in drops a letter from Lord John [Russell], to say he had recommended me to H.M. for that very office!" [1]

There will probably always be a certain degree of

* The political economist Professor Richard Jones (1790–1855).

mystery as to Herschel's motives in taking a step which was bound to give his life a wholly new and unknown course at the conclusion of a career that had already seen so much change. The possibility of financial need cannot be ruled out. Herschel must have known that this post was far removed from his scientific life and that it would rudely disturb his enjoyment of the domestic happiness that was so important to him.

Undoubtedly he was partly moved by a conviction that it was his moral duty to use his energy and capabilities to promote the public good of his country as well as his own inclinations and personal interests, even though public service would mean giving up his scientific activities. The modest view that he always took of his scientific achievements led him to overlook the basic fallacy of this belief. He failed to realize that he could render far greater service to his country as a scientist than he could in public office, purely administrative in nature and requiring, not scientific qualifications, but rather a knowledge of public finance and political economy and organizational ability.

The fact that the office of Master of the Mint had once before been held by a great scientist—Sir Isaac Newton himself—may possibly have given Jones the idea of procuring it for Herschel. In Newton's time, however, the Mastership of the Mint had been a purely honorific post involving chiefly representational duties. Afterward it developed into a political office, held by a member of the Cabinet with great influence on the financial policy of the Empire, until an Act of Parliament passed in December 1850, after the resignation of Herschel's predecessor, Shiel, made it a purely administrative position. The Mint Board was officially dissolved on March 15, 1851, and after the death of Sir James Morrison the post of Deputy Master, which ranked second to Master, was filled by Captain Harness of the Royal Engineer Corps, with whom Herschel was acquainted in an entirely different capacity. A number of privileges hitherto enjoyed by various employees of the Mint, and certain private contracts that had existed between them and the Mint authorities, were discontinued, and certain individuals, including Bingley, the Queen's Assay Master, suffered considerable financial losses as a result. Herschel, who was appointed Master in December 1850, was faced early in his administration with the difficult and thankless task of carrying out the reforms decided upon by Parliament, which involved drastic changes in the circumstances of many of the staff and aroused a wave of discontent. Thus Herschel entered upon his new duties under very unfavorable auspices.

The introduction of the staff reform was announced by Herschel in an office memorandum dated January 25, 1851: "All persons employed in the Mint are equally the servants of the Sovereign, and all will perform their duties under the immediate orders of the Master of the Mint." [2] This order had the effect of excluding every form of private work that had previously been customary. For example, the engravers who produced the molds for the coins and medals had been allowed to carry out private jobs using the machines and facilities of the Mint; this right was now abolished. In particularly hard cases Herschel made adjustments. William Wyon, the chief engraver, was permitted to fill a limited number of private contracts, though without using the Mint's staff or machines. On the whole, however, the staff reform struck a blow at old-established rights and customs.

Of greater biographical interest are Herschel's own plans for improvement, notably his efforts to change British currency to the decimal system. Herschel had been concerned with this problem in another form before he became associated with the Mint. During the period from 1838 to 1843, he had belonged to a Royal Commission set up to examine the possibility of introducing the metric system of weights and measures. The question of changing the currency to the decimal system had been raised in Parliament in 1841, but without result. When he was appointed Master of the Mint, Herschel became a strong advocate of the decimal system and was supported by the "Committee for superintending the construction of National Standards" to which his friends Airy, Lord Rosse, Peacock, and the astronomers Sir John William Lubbock (1803–1865) and Richard Sheepshanks (1794–1855) belonged. Herschel himself proposed to replace the pound sterling by a "100-millet" coin for which he suggested the name "Rupee" or "Rose." However, another Royal Commission, appointed in 1856, decided against the change and the old coinage continued in force.

Herschel did, however, introduce one technical innovation worth mentioning: he computed new tables of the purity of gold and silver coins and established accurate values for these. The trial plates (which Herschel called "fiducial pieces") used up to then were abolished at the same time.

Despite its difficulties, Herschel's performance as Master of the Mint was highly successful. His just and humane administration soon won the confidence both of his subordinates and of his superior Sir Charles Wood, Chancellor of the Exchequer. However, nothing he could do lessened the fierce opposition of those employees who had been hit especially hard by the staff reform, and in addition, he was overburdened with official work. Apart from his daily round of duty, he had to familiarize himself with the literature on public finance and political economy, and in the evenings he usually took with him on returning to his gloomy town residence in Harley Street a stack of official letters and files as homework.

As he had done all his life, he also took on other burdens. Since 1850 he had been a member of a Royal Commission appointed to consider the reform of the curriculum of studies and examinations at the University of Cambridge.

At the sessions of this Commission, he advocated with his usual energy the abolition of antiquated practices, partly of medieval origin, in favor of a modern and liberal policy. He even had a difference of opinion over this with his faithful but conservative friend Whewell, the Master of Trinity. As has been mentioned, Herschel declined the offer of membership of Parliament for Cambridge University.

During part of this period he served as a member of the Commission on scientific instruments for the Great Exhibition of 1851. Every afternoon, on leaving his office at the Mint, he hastened to the Exhibition stands, where he conducted negotiations with representatives of science and technology from all over the world, dealt with piles of letters and inquiries, and worked until the gates were closed, before going home to deal with Mint correspondence and files. Whatever he could not finish in the evening was dealt with early the next morning. Almost every day he rose at six a.m., worked until nine or ten, and then hurried off to the Mint.

Herschel's letters and diaries provide much information on his life in London. It was in harsh contrast to Collingwood, and the records reveal a Herschel very unlike the man of happier times—a man undergoing physical and mental suffering, governed by a bitter feeling that the last stage of his life had now begun under very different conditions from what he had hoped for and expected. He suffered above all by reason of his separation from his home and family, although frequently his wife or some of the children—most often the two eldest daughters, Caroline and Isabella—paid visits to Harley Street to dispel the loneliness of evenings and Sundays.

This gloomy life, however, was not entirely lacking in joyful family events. Among these were the marriage of his daughter Caroline to Alexander Gordon on December 9, 1852, and the brilliant success with which William

James Herschel, during the same month, graduated from Haileybury School, where he had been preparing for the Indian Civil Service. All the same, these were years of deprivation and sacrifice, and Herschel's scientific researches were dormant. Only on rare occasions did he find time to attend meetings of the Royal Society or the Royal Astronomical Society or to read their journals. "The meeting with men of science is now accompanied to me with feelings too painful to prolong more than absolute good breeding requires," he wrote in his diary on November 18, 1853.[3] Only rarely do his entries describe scientific events, such as annual visitations to Greenwich Observatory, or the visits of scientific friends, or contain hastily written reports on the activities of geomagnetic stations and the like. Serious observations were out of the question, though he noted in his diary on August 10, 1853: ". . . At night (which was very clear) I saw several meteors about Cassiopeia and the surrounding region (all the sky I could command) in a few minutes."[4]

It soon became evident that the load of work that Herschel had to undertake was undermining his health and equanimity to an ever-increasing extent. His diary contains several remarks suggestive of a serious crisis, although they do not make clear what the nature of the crisis was. It was provoked by a combination of physical ill health and nervous overstrain. Herschel's tendencies to bronchitis, rheumatism, and attacks of gout, which had resulted from the years of night watches at the telescope, caused him a great deal of trouble during his time in London. He spent many a sleepless night in efforts to ward off the pain with opium and other strong sedatives. His mind was troubled by the abandonment of his scientific researches and by his continued and increasing difficulties with the personnel of the Mint. With his mild nature, and his desire to prevent harm to anyone, Herschel was unable to assert himself. Indeed, his efforts at adjustment and his attempts to be fair to all the interests of his subordinates actually seem to have exacerbated the situation instead of calming it down.

In the depths of depression and despair, he wrote in his diary on August 17, 1853: "I desire here to record my deliberate opinion that any man having anything to do with the Public Service under the Treasury who can do anything else is a fool. Though it is writing my own condemnation." [5]

From the beginning of 1853, Herschel seems to have entertained the idea of resigning from the Mint. On May 17 of that year he had a long interview with Sir Charles Edward Trevelyan, the assistant secretary to the Treasury, in the course of which he expressed his views on the future administration of the Mint and mentioned his intention of resigning. Trevelyan apparently succeeded in dissuading him from doing so for the time being, and he continued in office, although working conditions did not improve and his state of health deteriorated rapidly. He tried all kinds of distractions, reading thrilling novels, frequenting society, and spending days of relaxation at Collingwood, but the cheerful atmosphere of the family circle made his own miserable situation stand out in even starker contrast when he returned to Harley Street.

Late in 1854 a nervous breakdown compelled him for a time to give up working at the Mint almost entirely. In March 1855, his gout became so bad that he had to return to Collingwood and spend several weeks in bed or in a wheelchair. He was sufficiently restored in September to take a trip to France with his family, but this did not bring about a complete cure. New symptoms developed— abscesses and attacks of fever. Herschel endured all his sufferings with patience. "What God sends is welcome," he wrote in his diary on November 18, 1855. [6]

At last, he was forced to carry out his intention of resigning. The Chancellor of the Exchequer let him go with great reluctance. It was difficult to find a successor with the capacity to continue the reforms that Herschel had so energetically begun and to match him in authority, sense of duty, and personal qualities. At last a man who appeared suitable was found in the person of the Scottish chemist, Thomas Graham (1805–1869), Professor of Chemistry at London University and a Fellow of the Royal Society. Graham had made a distinguished name for himself with his experiments on the absorption of gases and various technical inventions (especially in the purification of water supplies), and he combined organizing ability with deep scientific knowledge which he used to advantage in the production of coins. He succeeded Herschel on April 27, 1856.

After his resignation, Herschel led a secluded life at Collingwood. His recovery progressed, although his general state of health continued to be shaky. For the time being, major scientific enterprises were out of the question. His painful awareness of his numerous afflictions and of the gradual ebbing of his strength was mitigated by family affairs. The diary records marriages of his daughters, the birth of grandchildren, and the professional progress of his sons. Scientific friends called on Herschel in his quiet study and were always welcomed. He continued to be regarded as the unchallenged authority in England on all fields of mathematics and physics, and his advice was widely sought and freely given.

Little by little, Herschel returned to scientific research on a small scale, but the varied subject matter and the brevity of most of his papers indicate that he was mostly working on chance ideas with no systematic program. In many cases the papers were stimulated by his voluminous correspondence with scientific colleagues.

He resumed his experiments in physics and chemistry and was especially interested in following the progress of photography. A frequent guest at Collingwood was Julia

Margaret Cameron, one of the most distinguished photographic portraitists in England during the early days of photography. For her visits of several days, she always brought an entire traveling laboratory. To her, posterity owes several fine portraits of Herschel. One of these appears as the frontispiece of this book, and another was a source for the posthumous portrait by Herschel's daughter Caroline.

Herschel had realized at an early stage that photography was destined to make an immense contribution to research in astronomy. He had suggested in his *Results* (see Chapter 7) that events on the surface of the sun should be kept under surveillance with the aid of photography. His ideas on celestial photography are outlined in a short paper in the *Monthly Notices* of 1854–55.[7]

Herschel's work at the telescope was now confined to regular solar observations, counting sunspots and examining the granulation. He developed his own views on the nature of sunspots,[8] and discussed the solar theory of the Scottish engineer James Nasmyth (1808–1890) in a published letter.[9] This was the period when the foundations of solar physics and of astrophysics in general were being laid by Kirchhoff's and Bunsen's invention of spectrum analysis, a development which began while Herschel was still alive.

He continued to keep in touch with research in terrestrial magnetism. At the annual meeting of the British Association at Leeds in 1858, he reported on current geomagnetic and meteorological observations at the various stations of the British Empire [10] and gave the opening address at the meetings of the Chemistry Section,[11] where he served as president. The following year he acted as one of the vice-presidents of the Association during its meeting at Aberdeen.

The Aberdeen meeting was his last major public appearance. Thereafter he withdrew more and more into his private sphere, feeling that his active part in science was gradually coming to an end. He accepted this with the composure of an aging man, and acceptance was made easier by the fact that his mind, still active and creative as ever, had turned to an abundantly fruitful task which he regarded as the work of his old age. This was a general catalogue of the nebulae and star clusters that had been discovered and observed up to 1863. The catalogue appeared in the *Philosophical Transactions* in 1864 and includes 5079 objects (reduced to the equinox of 1860).[12] Since almost all the nebulae had been discovered by his father or himself, this catalogue is a product of the observational industry of two generations of astronomers unparalleled in the history of the subject. Nor was it merely a historical monument. The considerably extended version of it published by J. L. E. Dreyer in 1888 with the title "A New General Catalogue of Nebulae and Clusters" (the N.G.C.), which contains some 13,000 objects, still holds its place along with newer and more extensive compilations as an indispensable reference work at all observatories and astronomical institutions.[13]

Herschel spent several years on this catalogue, the completion of which demanded painstaking attention to detail. According to his own estimate, some 50,000 separate entries were needed in order to provide all the necessary data for the various objects. An entry in his diary on July 23, 1863, reads: "Finished . . . my nebulae catalogue. . . . It has been a long and laborious work.— Query now about the double stars—at all events the collection into one work of *all* my father's measures of positions and distances." [14] This thought resulted in a paper in the *Memoirs of the Royal Astronomical Society* for 1867.[15]

As the diary indicates, Herschel was also contemplating a much greater plan. As a counterpart to the nebulae catalogue, he intended to produce a general list of all double stars observed and measured to date, giving for each system all the available measurements of position angle and separation so as to make it possible to determine a large number of orbits.

To this final catalogue Herschel devoted the last seven years of his life, though the period was clouded by numerous illnesses and overshadowed by a noticeable decline in vigor. With immense persistence, he worked on the catalogue almost daily. It finally included 10,300 double stars. The catalogue itself, with some additions by the Oxford astronomer Robert Main (1808–1878) and Pritchard, was published after his death in the *Memoirs of the Royal Astronomical Society*,[16] but the "General History of Double Stars," in which Herschel gave a detailed discussion and listing of all the available data for some 5000 stars, was never published. The manuscript, duplicated quarto sheets of tables in twenty-four cardboard boxes, is in the possession of the Royal Astronomical Society in London.

In his last years Herschel engaged in a number of projects in addition to his astronomical catalogues. His articles for the eighth edition of the *Encyclopaedia Britannica* have already been mentioned. The repeated new editions of *Outlines of Astronomy* (six had appeared by 1863) he prepared with the greatest care. Twenty-five contributions to scientific journals and periodicals were written during the last ten years of his life.[17] The diversity of their subject matter—one was a major treatise "On Musical Scales"[18] bears impressive testimony to the supreme mastery over the whole range of physical science still possessed by their aged author.

Apart from his works addressed to readers with scientific training, Herschel never considered it beneath him to address a wider public in elementary essays couched in popular terms for the purpose of spreading scientific knowledge. He wrote articles on topics in astronomy, physics, and geology for the family magazine *Good Words*, which had a large circulation. He gave lectures in the same vein ("About Volcanoes and Earthquakes," "The

Sun," "On Comets") at the schoolhouse of the village of Hawkhurst. It gave him profound satisfaction to unlock the results of scientific research to those sections of the community from whom they would otherwise have remained hidden. This is further evidence of his interest in teaching and his appreciation of the importance of a wide-ranging popular education. Public education was not taken for granted then as it is now, though beginnings were being made in the major cities. Numerous museums were being founded, and societies such as the Royal Institution in London arranged educational lectures for the general public. But in a village like Hawkhurst, with only a few hundred inhabitants, an enterprise of this kind, undertaken by a man of Herschel's stature, was extraordinary and unique. More than most scholars of his time, Herschel was imbued with the idea of making science accessible to people without specialized academic knowledge. This educational purpose is expressed in many of his scientific works, notably his encyclopedia articles and the *Outlines*, and in a sense also in the *Preliminary Discourse*.

Herschel collected his popular lectures and essays in *Familiar Lectures on Scientific Subjects*.[19] Though their content is now quite out of date, they still make pleasant reading and give a stimulating presentation of scientific problems. They are both easy to understand and meticulously accurate as to scientific facts.

Another book has a special charm because it reflects an appealing aspect of Herschel's personality: his love for poetry. This is his translation of Homer's *Iliad*, which was made during the few hours of relaxation he allowed himself from his scientific work. The project came about by chance. An article in *The Times* (London) about an English translation of Homer gave Herschel the idea of making a translation of the *Iliad* in English hexameters, a meter to which the language is not especially well adapted. He determined to adhere strictly to the meaning of the original, in contrast to the poet Alexander Pope's eighteenth-century translations of Homer. Herschel even went so far as to use a special kind of type to distinguish words which were not in the Greek text. According to Whewell, it was amazing how few such words he had to insert.

Part of the inspiration seems to have come from the German translation of Homer by Johann Heinrich Voss, which let loose a wave of enthusiasm for Homer in Germany. Herschel worked on his translation for four years. He finished it on November 8, 1865, and it was published in 1866.[20]

The work had a mixed reception. The poet Alfred Tennyson called it a "burlesque barbarous experiment," but Whewell, himself an enthusiastic protagonist of English hexameters, preferred Herschel's to any other English version. However great or small its literary and artistic merits may be, the "Collingwood *Iliad*," as A. M. Clerke called it, certainly provides a most engaging contrast to the serious scientific works of Herschel's old age.

When he finished his translation of the *Iliad*, Herschel was in his seventy-fourth year. Despite his physical weakness, his mental powers were unimpaired and he was able to achieve all the scientific goals he had set for himself in the last stage of his life. His family life was only once clouded by a tragic event: his daughter Margaret Louisa died in 1861 at the age of twenty-seven, shortly after her marriage. His sons William James and John were both in Government service in India, the first as an official of the Colonial Administration and the other as an engineer in the Royal Bengal Engineering Corps, which was commissioned to carry out geodetic surveys in Bengal. John was a keen astronomer in addition to his professional duties; he carried out spectroscopic studies of the southern sky and was elected to the Royal Society during his father's lifetime. The third son, Alexander Stewart, took up an academic career after completing his studies at Cambridge and was appointed to a chair of astronomy and physics at Glasgow. He too became a Fellow of the Royal Society and made a distinguished name for himself in research on meteors. He discovered the cometary origin of a number of meteor streams and ranks as one of the founders of modern meteoritics. With some sixty scientific contributions to periodicals, he is creditably represented next to his father in the Royal Society's *Catalogue of Scientific Papers.*[21]

John Herschel continued to carry on a lively correspondence with scientific friends at home and abroad until the very last year of his life. He rarely left Collingwood, where he worked on his catalogue of double stars every day, carried out chemical and photographic experiments, and even made occasional astronomical observations, though his physical strength declined steadily. Time and again the entries in his diary mention illnesses and ailments of various kinds. Attacks of gout alternated with serious bouts of bronchitis; for days on end he could hardly move or do so only with great pain, and often he was confined to his wheelchair. References to death began to appear in the diary. When his old friend William Whewell died on March 6, 1866,[22] he wrote: "Our dying friends come o'er us like a cloud to damp our brainless ardour and take off that glare of life that often blinds the view. But life has no glare to a man entering on his 75th year. It is the shade and the calm that he longs for." [23]

When his friend Augustus De Morgan died in the spring of 1871, Herschel wrote to the widow: "Many and very distinct indications tell me that I shall not be long after him." [24]

His premonition was to be fulfilled. On May 11, 1871, John Herschel died in his house at Collingwood after a final short illness, at the age of seventy-nine.

A wave of grief and sympathy swept through England

and the scientific world. He was mourned not only because of the loss of a man whose life had been so richly endowed with both intellectual and human values, but also for the passing of the era that ended with his death. It had been an era of intellectual universality and of automatic awareness of the deeper philosophical significance of scientific efforts; it had been marked by the realization that such efforts could not be allowed to be merely a means to an end, but that they involved profound ethical values. Herschel was by no means the only man of his time to embody the ideal of universal learning—there were many others both in England and elsewhere—but he was the one who embodied it most successfully.

In an obituary of Herschel, a contemporary used the words *Non tetigit, quod non ornavit* [He touched nothing that he did not adorn]." [25] In reference to Herschel, the meaning of this phrase can be extended: a significant part of the brilliance of his researches arose from the fact that his whole personality was involved. The warmth of his humanity, so richly enjoyed by his contemporaries, can still be appreciated in a colder and more practical age.

REFERENCE NOTES

BIBLIOGRAPHY

INDEX

Reference Notes

CHAPTER 1
THE YOUNG PRODIGY

1. "J. F. W. Herschel," *Monthly Notices of the Royal Astronomical Society*, vol. 32, 1872, no. 4, p. 122.

2. *Philosophical Transactions of the Royal Society*, 1781, pp. 492–501. [Hereinafter referred to as *Philosophical Transactions*]

3. Agnes M. Clerke, *A Popular History of Astronomy during the Nineteenth Century*, 4th ed. (London, A. & C. Black, 1902), p. 78.

4. *Memoir and Correspondence of Caroline Herschel* (1750–1848), edited by Mrs. John Herschel (London, John Murray, New York: Appleton & Co., 1876), p. 259.

5. Isaac Newton, *Philosophiae Naturalis Principia Mathematica*. London, 1687.

6. Robert Woodhouse, *A Treatise on Plane and Spherical Trigonometry*. London, 1809.

7. S. F. Lacroix, *An Elementary Treatise on the Differential and Integral Calculus*. Translated [Part 2 by G. Peacock and J. F. W. Herschel]; with an appendix [by J. F. W. Herschel] and notes [by G. Peacock and J. F. W. Herschel]. Cambridge, 1816.

8. J. F. W. Herschel, *A Collection of Examples of the Application of Calculus of Finite Differences* . . . (Cambridge: Deighton, 1820).

9. "Analytical formulae for the tangent, cotangent, etc.," *Nicholson's Journal,* XXXI, 1812, p. 133.
10. "Trigonometrical formulae for sines and cosines," *ibid.,* XXXII, 1812, p. 13.
11. "On a remarkable application of Cotes's Theorem" [1812], *Philosophical Transactions,* 1813, pp. 8–26.
12. "Considerations of various points of analysis," *Philosophical Transactions,* 1814, pp. 440–468; "On the development of exponential functions . . . ," *ibid.,* 1816, pp. 25–45; "On circulating functions . . . ," *ibid.,* 1818, pp. 144–168.
13. A manuscript copy of the "Statutes of the Analytical Society" (presumably an early draft) is bound in with the copy of Vol. 1 of the *Memoirs of the Analytical Society* in the library of St. John's College, Cambridge.
14. "On equations of differences and their application to the determination of functions from given conditions," in 3 parts, *Memoirs of the Analytical Society,* 1813, pp. 65–114.
15. Letter dated February 8, 1813; copy in the possession of the Royal Society, London.
16. Letter dated October 13, 1813; copy in the possession of the Royal Society.
17. Letter dated January 14, 1814; copy in the possession of the Royal Society.
18. Letter dated September 20, 1814; copy in the possession of the Royal Society.
19. Letter dated March 23, 1815; copy in the possession of the Royal Society.
20. Letter dated November 6, 1815; copy in the possession of the Royal Society.
21. Letter dated December 18, 1815; copy in the possession of the Royal Society.
22. Letter dated October 10, 1816; copy in the possession of the Royal Society.

CHAPTER 2
VERSATILITY, VOCATION, AND TRAVEL

1. "On the action of crystallized bodies on homogeneous light" . . . [1819], *Philosophical Transactions,* 1820, pp. 45–100.
2. William Thomson, Lord Kelvin, *Report of the 41st Meeting of the British Association . . . held at Edinburgh in August, 1871,* p. 86.

3. "On a remarkable peculiarity in the law of extraordinary refraction of differently-colored rays, exhibited by certain varieties of Apophyllite" [1821], *Transactions of the Cambridge Philosophical Society*, I, 1822, pp. 241–248.

4. "On the absorption of light by coloured media, and on the colours of the prismatic spectrum exhibited by certain flames . . ." [1822], *Transactions of the Royal Society of Edinburgh*, IX, 1823, pp. 445–460.

5. Gustav Robert Kirchhoff, "Untersuchungen über das Sonnenspektrum und das Spektrum chemischer Elemente [Investigations of the Solar Spectrum and the Spectrum of Chemical Elements]," *Publications of the Royal Prussian Academy of Sciences*, 1861.

6. See William Herschel, "Observations tending to investigate the nature of the sun . . . ," *Philosophical Transactions*, 1801, pp. 265–318.

7. "On the optical phenomena exhibited by mother-of-pearl, depending on its internal structure," *Edinburgh Philosophical Journal*, II, 1820, pp. 114–221.

8. *Philosophical Magazine*, series 3, III, 1833. Quoted in Edmund Hoppe, *Geschichte der Physik* [History of Physics], (Brunswick, Germany, 1926), p. 147.

9. "On the aberration of compound lenses and object glasses," *Philosophical Transactions*, 1821, pp. 222–267.

10. "Rules for the determination of the radii of a double achromatic object glass," *Edinburgh Philosophical Journal*, VI, 1822, pp. 361–370.

11. "On the hyposulphurous acid and its compounds," *Edinburgh Philosophical Journal*, I, 1819, pp. 8–29; "Additional facts relative to the hyposulphurous acid," *ibid.*, pp. 396–400.

12. See J. F. W. Herschel, "Mathematics," *Edinburgh Encyclopaedia*, 1830, vol. 13, pp. 359–383; "Isoperimetrical Problems," *ibid.*, vol. 12, pp. 320–328.

13. Letter dated April 27, 1818; copy in the possession of the Royal Society.

14. "On a new method of computing occultations of the fixed stars," *Memoirs of the Astronomical Society*, I, 1822, pp. 325–328.

15. "Subsidiary tables for facilitating the computation of annual tables of the apparent places of 46 principal fixed stars . . . ," *ibid.*, pp. 421–496.

16. Undated letter written in the spring of 1819; copy in the possession of the Royal Society.

17. William Herschel, "Account of the changes that have happened, during the last 25 years, in the relative situation of the double stars, with an investigation of the cause to which they are owing," *Philosophical Transactions*, 1803, pp. 339–382.

18. J. F. W. Herschel and J. South, "Observations of the apparent distances and positions of 380 double and triple stars, made in the years 1821, 1822, and 1823, and compared with those of other astronomers . . . ," *Philosophical Transactions*, 1824 (pt. 3), 1–412.

19. James South, "Observations of the apparent distances and positions of 458 double and triple stars, made in the years 1823–1825 . . . ," *Philosophical Transactions*, 1826, pp. 1–391.

20. *Philosophical Transactions*, 1824, pp. 162–196.

21. *Edinburgh Journal of Science*, II, 1825, pp. 193–199.

22. *The Scientific Papers of Sir William Herschel*, compiled by J. L. E. Dreyer (2 vols., London: Royal Society and Royal Astronomical Society, 1912), vol. I.

23. J. F. W. Herschel and Charles Babbage, "Account of the repetition of M. Arago's experiments on the magnetism manifested by various substances during the act of rotation," *Philosophical Transactions*, 1825, pp. 467–496.

24. "On the separation of iron from other metals," *Philosophical Transactions*, 1821, pp. 293–299.

25. Letter dated April 12, 1820; copy in the possession of the Royal Society.

26. Reproduced in the manuscript of an unfinished and unpublished biography of J. F. W. Herschel by Mira F. Hardcastle, in the possession of Mrs. E. D. Shorland of Bracknell, Berkshire, England; hereinafter referred to as "Hardcastle ms."

27. J. F. W. Herschel and Charles Babbage, "Barometrical Observations Made at the Fall of Staubbach," *Edinburgh Philosophical Journal*, VI, 1821–1822, pp. 224–227.

28. Letter dated April 7, 1824; Hardcastle ms., "Grand Tour," p. 2.

29. Letter to Lady Herschel dated May 12, 1824; Hardcastle ms., "Grand Tour," p. 16.

30. *Memoir and Correspondence of Caroline Herschel, op. cit.,* p. 173.

31. Letter to Lady Herschel dated June 17, 1824; Hardcastle ms., "Grand Tour," p. 27.

32. *Memoir and Correspondence of Caroline Herschel, op. cit.,* p. 173.
33. Letter dated August 24, 1824; Hardcastle ms., "Grand Tour," p. 60.
34. *Wilh. Herschels sämtliche Schriften. Herausgeben von J. W. Pfaff. Bd. 1: Über den Bau des Himmels.* Dresden and Leipzig: Arnold, 1826.
35. Diary entry on October 27, 1824; Hardcastle ms., p. 272.
36. Letter to Lady Herschel dated September 11, 1826; Hardcastle ms., p. 296.
37. Letter to Caroline Herschel dated September 17, 1826, *Memoir and Correspondence of Caroline Herschel, op. cit.,* p. 203.
38. Sir William Rowan Hamilton to J. F. W. Herschel, October 12, 1827: Hardcastle ms., p. 333.

CHAPTER 3
THE ASTRONOMICAL HERITAGE

1. "Observations of nebulae and clusters, made at Slough, with a 20-feet reflector, between the years 1825–1833," *Philosophical Transactions,* 1833, pp. 359–506.
2. "Descriptions and approximate places of 321 new double and triple stars," *Memoirs of the Astronomical Society,* ii, 1826, pp. 459–497; "On the approximate places and descriptions of 295 new double and triple stars . . .," *ibid.,* iii, 1829, pp. 47–63; "Third series of observations . . . containing a catalogue of 384 new double and multiple stars . . .," *ibid.,* pp. 177–214; "Fourth series . . . containing the mean places and other particulars of 1236 double stars . . .," *ibid.,* iv, 1831, pp. 331–378; "Fifth catalogue of double stars . . . containing the places, descriptions, and measured angles of positions of 2007 of these objects . . .," *ibid.,* vi, 1833, pp. 1–74; "Sixth catalogue of double stars . . . containing the places . . . of 285 of these objects . . .," *ibid.,* ix, 1836, pp. 193–204.
3. *Memoir and Correspondence of Caroline Herschel, op. cit.,* p. 188.
4. Caroline Herschel, "Reduction and arrangement in the form of a catalogue in zones of all the star clusters and nebulae observed by Sir William Herschel," *Memoirs of the Astronomical Society,* iii, 1829.
5. *Philosophical Transactions,* 1833, p. 361.
6. Letter to Caroline Herschel written in May 1827; Hardcastle

ms., p. 320; reproduced in *Memoir and Correspondence of Caroline Herschel, op. cit.,* p. 213.

7. Letter dated February 12, 1828; copy in the possession of the Royal Society, London.

8. Charles Messier, *Catalogue des Nébuleuses . . ., Connaissance des Temps,* 1784.

9. "An account of the actual state of the Great Nebula in Orion, compared with those of former astronomers"; Observations of the Nebula in the Girdle of Andromeda" [with other observations], *Memoirs of the Astronomical Society,* II, 1826, pp. 459–497.

10. See Agnes M. Clerke, *The Herschels and Modern Astronomy* (London: Cassell, 1895), p. 154.

11. "On the investigation of the orbits of revolving double stars," *Memoirs of the Astronomical Society,* v, 1833, pp. 171–222.

12. See *Proceedings of the American Philosophical Society,* vol. 12, 1871, p. 218.

13. *Philosophical Transactions,* 1826, pp. 256–280.

14. *Ibid.,* pp. 266 f.

15. *Ibid.,* pp. 267 f.

16. Friedrich Wilhelm Bessel, "Bestimmung der Entfernung der 61 Sterne des Schwans ["Determination of the Distance of 61 Stars in Cygnus]," *Astronomische Nachrichten,* 1840.

17. *Memoir and Correspondence of Caroline Herschel, op. cit.,* p. 194.

18. Letter to Francis Baily, written on February 12 or March 12, 1828; copy in the possession of the Royal Society, London.

19. Letter dated January 3, 1827; copy in the possession of the Royal Society.

20. Letter written in the summer or autumn of 1827; copy in the possession of the Royal Society.

21. Letter dated October 17, 1826; copy in the possession of the Royal Society.

22. J. F. W. Herschel, *Preliminary Discourse on the Study of Natural Philosophy.* New ed. (London: Longmans, 1830), p. 361. [Hereinafter referred to as *Preliminary Discourse. . . .*]

23. Charles Darwin, *The autobiography 1809–1882 . . .,* ed. by Nora Barlow (London, 1958), pp. 67f.

24. *Preliminary Discourse . . ., op. cit.,* p. 15.

25. *Ibid.,* p. 6.

26. *Ibid.,* p. 7.

27. *Ibid.,* p. 10.

28. J. F. W. Herschel, *A Treatise on Astronomy*. London: Longmans, 1833.
29. J. F. W. Herschel, "Light," *Encyclopaedia Metropolitana* (London, 1845), vol. 4, pp. 341–586.
30. Letter dated January 14, 1829; Hardcastle ms., next to p. 284.
31. "Light," *op. cit.*, p. 450.
32. *Ibid.*
33. J. F. W. Herschel, "Sound," *Encyclopaedia Metropolitana*, *op. cit.*, vol. 4, pp. 763–824.
34. "Account of a series of observations made in the summer of 1825, for the purpose of determining the difference of meridians of the Royal Observatories of Greenwich and Paris," *Philosophical Transactions*, 1826, pp. 266–280.
35. Diary entry, September 22, 1828; Hardcastle ms., p. 336.
36. James Grahame to Patrick Stewart, November 29, 1828; Hardcastle ms., p. 340.
37. M. B. Stewart to J. F. W. Herschel (probably December 1828 or January 1829); Hardcastle ms., p. 339.
38. Agnes M. Clerke, "Sir John Frederick William Herschel," *Dictionary of National Biography* (London, 1891), vol. 26, p. 264.
39. Hardcastle ms., pp. 343 ff.
40. *Memoir and Correspondence of Caroline Herschel, op. cit.*, p. 239.
41. Letter dated May 23, 1832; copy in the possession of the Royal Society.
42. *Memoir and Correspondence of Caroline Herschel, op. cit.*, pp. 254 f.
43. *Ibid.*, p. 260. The paper "On the satellites of Uranus" [1834] was published in *Memoirs of the Astronomical Society*, VIII, 1835, pp. 1–24.

CHAPTER 4

AT THE CAPE OF GOOD HOPE

1. Diary entry, January 1, 1834; copy in the possession of the Royal Society.
2. J. F. W. Herschel, *Results of Astronomical Observations made during the Years 1834, 5, 6, 7, 8 at the Cape of Good Hope, being a completion of a telescope survey of the whole surface of the visible heavens commenced in 1825* (London: Smith,

Elder & Co., 1847), p. xvi. [Hereinafter referred to as *Results. . . .*]

3. *Ibid.*, p. vii.
4. See Clerke, *The Herschels . . ., op. cit.*, p. 173.
5. *Results . . ., op. cit.*, p. 381.
6. *Ibid.*, p. 380.
7. *Philosophical Transactions*, 1796, pp. 166–226; *ibid.*, pp. 452–482; *ibid.*, 1797, pp. 293–324; *ibid.*, 1799, pp. 121–144.
8. Diary entry, March 5, 1836; copy in the possession of the Royal Society.
9. *Results . . .*, p. 146.
10. *Ibid.*, p. 147.
11. *Ibid.*, p. 34.
12. *Ibid.*, p. 37.
13. *Ibid.*, p. 25.
14. *Ibid.*, p. 31.
15. J. F. W. Herschel, *Outlines of Astronomy*, 2nd ed. (London: Longmans, 1849), p. 598.
16. William Herschel, "On nebulous stars, properly so called," *Philosophical Transactions*, 1791, p. 83.
17. Diary entry, October 28, 1835; copy in the possession of the Royal Society.
18. *Results . . ., op. cit.*, p. 409.
19. *Ibid.*, p. 434.
20. *Ibid.*, p. 435 *n.*
21. *Ibid.*, Appendix, p. 445.
22. *Ibid.*, Appendix, p. 446.
23. In the possession of Mrs. E. D. Shorland.
24. Mostly in the possession of the South African Public Library, Cape Town.
25. Victor Cousin, *Report on the state of public instruction in Prussia, with plans of school houses,* translated by Austin (London, 1834).
26. Letter to John Bell, dated February 17, 1838. See also W. T. Ferguson and R. F. M. Immelman, *Sir John Herschel and Education at the Cape, 1834–40* (Cape Town: Oxford University Press, 1961), p. 21.
27. Memorandum from Herschel to Lord Glenelg dated March 6, 1838, Ferguson and Immelman, *op. cit.*, p. 28.
28. Letter to Napier written in November 1838; copy by Mira F. Hardcastle from copies of J. H.'s letters in the possession of the Royal Society. [Hereinafter referred to as Hardcastle extracts].

29. Clerke, *The Herschels* . . ., *op. cit.*, p. 181.
30. *Herschel at the Cape*, edited by David S. Evans *et al.* (Austin, Texas, and London: University of Texas Press, 1969).
31. Clerke, *The Herschels* . . ., *op. cit.*, p. 45.
32. Letter to Napier written in November 1838; Hardcastle extracts.
33. *Neueste Berichte vom Cap der Guten Hoffnung über Sir John Herschels höchst merkwürdige astronomische Entdeckungen den Mond und seine Bewohner betreffend.* Hamburg: Erie, 1836.
34. Letter to W. H. Sykes dated April 12, 1839; Hardcastle extracts.

CHAPTER 5
CONSTELLATION REFORM AND TERRESTRIAL MAGNETISM

1. Diary entry, April 21, 1838; copy in the possession of the Royal Society.
2. Diary entry, June 6, 1838; copy in the possession of the Royal Society.
3. *See also* "On the advantages to be attained by a revision and rearrangement of the constellations, with especial reference to those of the southern hemisphere, and on the principles upon which such re-arrangement ought to be constructed," *Monthly Notices of the Royal Astronomical Society*, v, 1839–43, pp. 116–118.
4. "Report of a Committee for revising the nomenclature of the stars," *British Association Reports*, 1844, pp. 32–42.
5. Jérome de Lalande, *A Catalogue of Those Stars in the* Histoire céleste française, *for Which Tables of Reduction to the Epoch 1800 Have Been Published by Prof. Schumacher.* Reduced at the expense of the *British Association* . . . under the immediate superintendence of the late Francis Baily (London, 1847).
6. Abbé Nicolas Louis de Lacaille, *A Catalogue of 9766 Stars in the Southern Hemisphere, for the Beginning of the Year 1750, from the Observations, Made at the Cape of Good Hope in the Years 1751 and 1752.* With a preface by Sir J. F. W. Herschel. London, 1847.
7. "Instructions for making and registering meteorological observations in Southern Africa, and other countries in the South Seas, and also at sea." Drawn up for circulation by the Meteorological Committee of the South African Literary and Philosophical Institution (reprint for private distribution, London, 1835).
8. "Report of a committee appointed for the purpose of super-

intending the scientific cooperation of the British Association in the system of simultaneous observations in terrestrial magnetism and meteorology," *British Association Reports,* 1841, pp. 38–41. Subsequent reports of the committee were published in *British Association Reports,* 1842, pp. 1–11; *ibid.,* 1843, pp. 54–60; *ibid.,* 1844, pp. 143–155; *ibid.,* 1845, pp. 1–73.

9. Letter dated December 3, 1838; copy in the possession of the Royal Society.

10. *Ibid.*

11. Diary entry, October 15, 1838; copy in the possession of the Royal Society.

12. Diary entry, November 6, 1838; copy in the possession of the Royal Society.

13. *British Association Reports,* 1844, *op. cit.*

14. Letter dated September 11, 1838; copy in the possession of the Royal Society.

CHAPTER 6

PHOTOGRAPHY AND PHOTOCHEMISTRY

1. *Proceedings of the Royal Society,* IV, 1839, p. 131.

2. "Chemical Experiments," vol. 3, manuscript notebooks of Herschel, in the possession of the Science Museum, London.

3. W. H. Fox Talbot, "Some account of the art of photogenic drawing . . .," *The Athenaeum,* February 9, 1839.

4. "Note on the art of photography, or the application of the chemical rays of light to the purpose of pictorial representation," *Proceedings of the Royal Society,* IV, 1839, pp. 131–133.

5. See Helmut and Alison Gernsheim, *The History of Photography* (New York: Oxford University Press, 1955), p. 81.

6. "Chemical Experiments," *op. cit.*

7. *Philosophical Transactions,* 1840, p. 2.

8. "Chemical Experiments," *op. cit.*

9. Gernsheim, *op. cit.,* p. 82.

10. Diary entry, February 13, 1839; copy in the possession of the Royal Society.

11. Gernsheim, *loc. cit.*

12. *Philosophical Transactions,* 1840, p. 3.

13. "Chemical Experiments," *op. cit.*

14. *Philosophical Transactions,* 1840, p. 12.

15. See Robert Hunt, *A Manual of Photography,* 4th ed. (London and Glasgow, 1854), pp. 221 f.

16. *Philosophical Transactions*, 1842, pp. 181–214.
17. "Contributions to actino-chemistry. On the amphitype, a new photographic process," *British Association Reports*, 1844 (pt. 2), pp. 12–13.
18. "On the chemical action of the rays of the solar spectrum on preparations of silver and other substances, both metallic and non-metallic, and on some photographic processes," *Philosophical Transactions*, 1840, pp. 1–60.
19. "On the action of the rays of the solar spectrum on vegetable colours, and on some new photographic processes," *Philosophical Transactions*, 1842, pp. 181–214.
20. Diary entry, February 13, 1839; copy in the possession of the Royal Society.
21. *Philosophical Transactions*, 1840, p. 19.
22. *Ibid.*, p. 24.
23. William Herschel, "Investigation of the powers of the prismatic colours to heat and illuminate objects" [with two supplements], *Philosophical Transactions*, 1800, pp. 255–326.
24. *Philosophical Transactions*, 1842, p. 186.
25. *Ibid.*, pp. 195 f.
26. See Jos. Maria Eder, *Photochemie* . . . (3rd ed., Halle, 1906); (*Ausführliche Handbuch der Photographie*, Vol. 1, Part 2).

CHAPTER 7
THE COLLINGWOOD PERIOD

1. Letter to E. Cooper dated September 23, 1838; Hardcastle extracts.
2. J. F. W. Herschel, "On the variability and the periodical nature of the star Alpha Orionis," *Memoir of the Royal Astronomical Society*, XI, 1840, pp. 269–278; *Monthly Notices of the Royal Astronomical Society*, V, 1839–43, pp. 11–16.
3. Letter dated January 19, 1839; Hardcastle extracts.
4. Letter dated March 30, 1839; Hardcastle extracts.
5. Letter dated July 22, 1844; Hardcastle extracts.
6. Diary entry, March 7, 1847; copy in the possession of the Royal Society.
7. Diary entry, March 17, 1843; copy in the possession of the Royal Society.
8. *The Athenaeum*, October 3, 1846, no. 988.
9. Letter dated December 27, 1846; Hardcastle extracts.
10. Most of Herschel's book reviews, speeches, and obituaries were later reprinted in *Essays from the Edinburgh and Quarterly*

Reviews, with Addresses and Other Pieces (London: Longmans, 1857). [Hereinafter referred to as *Essays*. . . .]

11. "Physical Astronomy," *Encyclopaedia Metropolitana*, vol. 3, pp. 647–729.
12. *Outlines of Astronomy, op. cit.*, p. v ff.
13. Clerke, *Dictionary of National Biography, op. cit.*, p. 264.
14. *Outlines of Astronomy, op. cit.*, p. 597.
15. *Ibid.*
16. Letter written in January 1845; Hardcastle extracts.
17. *A Manual of Scientific Enquiry; Prepared for the Use of Her Majesty's Navy and Adapted for Travellers in General.* Edited by Sir J. F. W. Herschel. Published by authority of the Lords Commissioners of the Admiralty. London: Murray, 1849.
18. *Encyclopaedia Britannica*, 8th ed., vol. 14, pp. 636–690.
19. J. F. W. Herschel, *Meteorology*, from the *Encyclopaedia Britannica*. Edinburgh: Black, 1861.
20. J. F. W. Herschel, *Familiar Lectures on Scientific Subjects*. London: Strahan, 1868.
21. *Encyclopaedia Britannica*, 8th ed., vol. 17, pp. 569–647.
22. *Ibid.*, vol. 21, pp. 117–145.
23. ἀμόρφωτα. No. I. On a case of superficial colour presented by a homogeneous liquid internally colourless"; "No. II. On the epipolic dispersion of light, being a supplement to No. I.," *Philosophical Transactions*, 1845, pp. 143–146, 147–154.
24. G. G. Stokes, "On the change of the refrangibility of light," *Philosophical Transactions*, 1852, pp. 463–562.
25. Clerke, *The Herschels . . ., op. cit.*, pp. 188 f.
26. *Ibid.*, p. 193.
27. Owned by Mrs. E. Shorland.
28. Friedrich von Schiller, *The Walk [Der Spaziergang]. Translated* [by Sir J. F. W. Herschel]. [1842.]
29. Herschel, *Essays . . ., op. cit.*
30. *Ibid.*, p. 741.
31. Diary entry, February 21, 1841; copy in the possession of the Royal Society.
32. Constance A. Lubbock, *The Herschel Chronicle: The Life Story of William Herschel and His Sister Caroline Herschel Edited by His Granddaughter Constance A. Lubbock*. Cambridge: The University Press, 1933.
33. Constance A. Lubbock, *A Short Biography of Sir John F. W. Herschel, Bt.* Cambridge [c. 1938].
34. Clerke, *Dictionary of National Biography, op. cit.*, p. 267.

CHAPTER 8
PUBLIC OFFICE AND LAST YEARS

1. Letter dated January 31, 1855; Hardcastle extracts.
2. See John Craig, *The Mint* (Cambridge: The University Press, 1953), p. 320.
3. Diary entry, November 18, 1853; copy in the possession of the Royal Society.
4. Diary entry, August 10, 1853; copy in the possession of the Royal Society.
5. Diary entry, August 17, 1853; copy in the possession of the Royal Society.
6. Diary entry, November 18, 1855; copy in the possession of the Royal Society.
7. "On the application of photography to astronomical observations," *Monthly Notices of the Royal Astronomical Society*, xv, 1854–55, pp. 158–159.
8. "On the solar spots," *Quarterly Journal of Science*, I, 1864, pp. 219–235.
9. "Letter respecting Mr. Nasmyth's 'willow leaves,'" *Monthly Notices of the Royal Astronomical Society*, xxv, 1865, pp. 152–153.
10. "Report of the Joint Committee of the Royal Society and the British Association, for procuring a continuance of the magnetic and meteorological observatories," *British Association Reports*, 1858, pp. 295–305.
11. *British Association Reports*, 1858, Chemical Section, pp. 41–45.
12. "Catalogue of nebulae and clusters of stars" [1863], *Philosophical Transactions*, 1864, pp. 1–137.
13. J. L. E. Dreyer, "A new general catalogue of nebulae and clusters of stars, being the catalogue of the late Sir John Herschel, revised, corrected and enlarged" [with appendix], *Memoirs of the Royal Astronomical Society*, XLIX, 1888.
14. Diary entry, July 23, 1863; copy in the possession of the Royal Society.
15. "A synopsis of all Sir William Herschel's micrometrical measurements and estimated positions and distances of the double stars described by him, together with a catalogue of those stars in order of Right Ascension, for the epoch 1880, so far as they are capable of identification" [1866], *Memoirs of the Royal Astronomical Society*, xxxv, 1867, pp. 21–136.
16. "A catalogue of 10,300 multiple and double stars, arranged in

the order of Right Ascension, by the late Sir J. F. W. Herschel . . .," ed. by R. Main . . . and Ch. Pritchard . . ., London: *Memoirs of the Royal Astronomical Society,* XL, 1874.

17. See *Catalogue of Scientific Papers* . . . London: The Royal Society of London, 1869, vol. 3, p. 328, nos. 125–131; 1877, vol. 7, p. 965, nos. 133–151.

18. *Quarterly Journal of Science,* v, 1868, pp. 338–352.

19. *Op. cit.*

20. *The Iliad of Homer.* Translated into English accentuated hexameters by Sir J. F. W. Herschel. London: Macmillan, 1866.

21. *Catalogue of Scientific Papers* . . ., *op. cit.,* vol. 7, pp. 963–964, nos. 7–61; *ibid.* (Cambridge), 1916, vol. 15, pp. 798–799 [another 23 publications].

22. Obituary notice of William Whewell, D.D., *Proceedings of the Royal Society,* XVI, 1866.

23. Diary entry, March 6, 1866; copy in the possession of the Royal Society.

24. See Clerke, *The Herschels* . . ., *op. cit.,* p. 197.

25. Proceedings of the American Academy of Arts and Sciences, vol. 8, 1872, p. 468.

Bibliography

✦

✦

✦

SELECTED PUBLICATIONS BY SIR JOHN HERSCHEL
(Listed in chronological order under each heading.)

BOOKS

A Collection of Examples of the Application of the Calculus of Finite Differences. Cambridge: Deighton, 1820.

Preliminary Discourse on the Study of Natural Philosophy. London: Longmans, 1830. (*The Cabinet Cyclopaedia. Natural Philosophy*).

A Treatise on Astronomy. London: Longmans, 1833. (The *Cabinet Cyclopaedia. Natural Philosophy*).

Results of Astronomical Observations Made during the Years 1834, 5, 6, 7, 8 at the Cape of Good Hope, Being a Completion of a Telescopic Survey of the Whole Surface of the Visible Heavens, Commenced in 1825. London: Smith, Elder, 1847.

Outlines of Astronomy. London: Longmans, 1849.

Essays from the Edinburgh and Quarterly Reviews, with Addresses and Other Pieces. London: Longmans, 1857.

Familiar Lectures on Scientific Subjects. London: Strahan, 1868.

CONTRIBUTIONS TO ENCYCLOPEDIAS

Edinburgh Encyclopaedia, The. Conducted by David Brewster. Edinburgh, 1830.
"Isometrical Problems" [not signed], vol. 12, part 1, pp. 320–328.
"Mathematics," vol. 13, part 1, pp. 359–383.

Encyclopaedia Metropolitana. Edited by Edward Smedley. London, 1845.

"Physical Astronomy," vol. 3, pp. 647–729.

"Light," [dated: Slough, Dec. 12, 1827], vol. 4, pp. 341–586.

"Sound" [dated: Slough, Feb. 3, 1830], vol. 4, pp. 763–824.

Encyclopaedia Britannica. 8th ed. Edited by T. S. Traills. Edinburgh: Black, 1857–1860.

"Meteorology," vol. 14 (1857), pp. 636–690.

"Physical Geography," vol. 17 (1859), pp. 569–647.

"Telescope," vol. 21 (1860), pp. 117–145.

BOOKS TRANSLATED OR EDITED BY SIR JOHN HERSCHEL

Lacroix, S. F. *An Elementary Treatise on the Differential and Integral Calculus.* Translation. [Pt. 2 by G. Peacock and J. F. W. Herschel.] *With an appendix* [by J. F. W. Herschel] *and notes* [by G. Peacock and J. F. W. Herschel]. Cambridge, 1816.

A *Manual of Scientific Inquiry; Prepared for the Use of Her Majesty's Navy: and Adapted for Travellers in General.* Edited by Sir J. F. W. Herschel. Published by authority of the Lords Commissioners of the Admiralty. London: Murray, 1849.

Includes "Meteorology" by J. F. W. Herschel.

Homer. *The Iliad of Homer. Translated into English Accentuated Hexameters by Sir J. F. W. Herschel.* London: Macmillan, 1866.

SCIENTIFIC PAPERS: BIBLIOGRAPHIES

A number of Herschel's papers for scientific journals are cited in the Reference Notes. A complete list would be inappropriate here; Herschel scholars are referred to the following bibliographies.

Complete bibliography: *Catalogue of Scientific Papers* (1800–1863), compiled and published by the Royal Society of London, vol. 3, 1869, pp. 322–328, nos. 1–131; vol. 7, 1877, p. 965, nos. 133–151.

List compiled by Herschel himself in 1861: *Mathematical Monthly Magazine*, Cambridge, Mass., III, p. 220.

Additional list: *Biographisch-literarisches Handwörterbuch zur Geschichte der exakten Wissenschaften . . .*, Gesammelt von J. C. Poggendorf. Leipzig 1863, Bd. 1, Sp. 1089–1091; Leipzig 1898, Bd. 3, p. 622.

BIOGRAPHICAL WORKS ON SIR JOHN HERSCHEL

Clerke, Agnes M. *The Herschels and Modern Astronomy*. London: Cassell, 1895. Biography of John Herschel, pp. 903–909.

———. *A Popular History of Astronomy during the Nineteenth Century*. 4th ed. London: A. & C. Black, 1902. Biographical sketch of John Herschel, pp. 45–50.

———. "Sir J. F. W. Herschel," *Dictionary of National Biography*, vol. 26. London: Smith, Elder, 1891. [Signed A. M. C.]

Evans, David S., *et al.* (eds.). *Herschel at the Cape*. Austin, Texas, and London: University of Texas Press, 1969.
Herschel's diaries and correspondence, 1834–1838, with a biographical introduction and extensive notes.

Herschel, Mrs. John (ed.) *Memoir and Correspondence of Caroline Herschel* (1750–1848). London: John Murray; New York: Appleton, 1876.
Includes letters from John Herschel.

Holden, Edward S. "The Three Herschels," *The Century Illustrated Magazine*, vol. 30, 1885, pp. 179–185.

Lubbock, Constance A. *A Short Biography of Sir John F. W. Herschel, Bt.* Cambridge [privately printed, c. 1938].

"Maria Mitchell's Reminiscences of the Herschels," *The Century Illustrated Magazine*, vol. 38, 1889, pp. 903–909.

Proctor, Richard A. *Essays on Astronomy*. London: Longmans, 1872.
"Sir John Herschel" (reprinted from *English Mechanic*, May 19, 1871), pp. 1–7; "Sir John Herschel as a Theorist in Astronomy" (reprinted from *The St. Paul's Magazine*, June 1871), pp. 8–28.

"Sir John Herschel at the Cape, 1834–1838." Special issue of *Quarterly Bulletin of the South African Library*, vol. 12, no. 3, December 1957.
Contains Introduction by D. H. Varley; "Sir John Herschel, 1792–1871" by R. H. Stoy; "The Astronomical Work of Sir John Herschel at the Cape" by David S. Evans; "Sir John Herschel's Contribution to Educational Developments at the Cape of Good Hope" by E. G. Pells.

OBITUARY NOTICES

Académie Royale des Sciences, des Lettres et des Beaux-Arts de Belgique. *Annuaire . . .*, 1872. "Notices sur J. F. W. Herschel par [L. A. J.] Quetelet," pp. 161–199.

American Academy of Arts and Sciences. *Proceedings* . . ., vol. 8, 1872, pp. 461–471.

American Philosophical Society. *Proceedings* . . ., vol. 12, 1871. "Obituary Notice of Sir J. F. W. Herschel, Bt., by Henry W. Field. Read . . . Dec. 1, 1871," pp. 217–223.

British Association for the Advancement of Science. *Report of the 41st Meeting . . . Held at Edinburgh in Aug. 1871.* London: Murray, 1872. "Address of Sir Wm. Thomson, Pres.," pp. lxxxv-lxxxvi. Also published in *Nature*, Aug. 3, 1871, pp. 262–263. Herschel and others.

Royal Astronomical Society. *Monthly Notices* . . ., vol. 32, 1872. No. 4. "J. F. W. Herschel," pp. 122–142. Signed C. P. [Charles Pritchard].

Royal Society of London, The. *Proceedings* . . ., vol. 20, 1872, pp. xvii-xxiii. Signed T. R. R. [T. Romney Robinson].

Smithsonian Institution, Board of Regents. *Annual Report* . . ., 1871. "Memoir of Sir John F. W. Herschel" by N. S. Dodge, pp. 109–135.

Stanley, A. P. "Science and Religion. A Sermon Preached in Westminster Abbey on May 12, 1871, Being the Sunday Following the Funeral of Sir John Herschel," *Good Words for 1871*, pp. 453–459.

OTHER SOURCES

Baier, Wolfgang. *Quellendarstellungen zur Geschichte der Fotografie.* Halle: Fotokinoverlag, 1964.

Ball, W. W. R. *A Short Account of the History of Mathematics.* 4th ed. London: Macmillan, 1927.

Balmer, Heinz. *Beiträge zur Geschichte der Erkenntnis des Erdmagnetismus.* Aarau, Switzerland: Sauerländer, 1956.

Cannon, Walter F. "John Herschel and the Idea of Science," *Journal of the History of Ideas*, vol. 22, no. 2, April-June 1961, pp. 215–239.

———. "The Impact of Cultural Uniformitarianism," *Proceedings of the American Philosophical Society*, vol. 105, no. 3, June 1961, pp. 301–314.

Two letters from John Herschel to Charles Lyell, 1836–1837.

Craig, John. *The Mint: A History of the London Mint from A.D. 287 to 1948.* Cambridge: The University Press, 1953. Chapter 8, pp. 317–323.

Eder, Joseph Maria. "J. F. W. Herschel," *Photographie Corre-spondenz*, 1887, no. 316, pp. 7–9.

———. *Die chemischen Wirkungen des Lichts (Photochemie)* . . ., Halle: Knapp, 1891.

———. *Geschichte der Photographie*. 2nd ed. Halle: Knapp, 1892; 4th ed. 1932.

Ferguson, W. T., and Immelman, R. F. M. *Sir John Herschel and Education at the Cape, 1834–1840*. Cape Town: Oxford University Press, 1961.

Gernsheim, Helmut. "Talbot's and Herschel's Photographic Experiments in 1838," *Image*, vol. 8, no. 3, September 1959, pp. 133–137.

Gernsheim, Helmut and Alison. *The History of Photography from the Earliest Use of the Camera Obscura in the 14th Century up to 1914*. New York and London: Oxford University Press, 1955.

Hoppe, Edmund. *Geschichte der Optik. (Webers illustrierte Handbücher.)* Leipzig: Weber, 1926.

———. *Geschichte der Physik*. Brunswick: Vieweg, 1926.

Hunt, Robert. *A manual of photography*. 4th ed., rev. London and Glasgow: Griffin & Co., 1854.

Malherbe, Ernst G. *Education in South Africa (1652–1922)*. Cape Town and Johannesburg: Juta & Co., 1925.

McIntyre, Donald. "The Herschel Obelisk," *Quarterly Bulletin of the South African Library*, vol. 8, no. 3–4, March–June 1954, pp. 87–92.

Newhall, Beaumont. "Sir J. F. W. Herschel," *The Complete Photographer*, 1942, no. 30, pp. 1963–1965.

Schiendl, C. *Geschichte der Photographie*. Vienna: Hartleben, 1891.

Schultze, R. S. "Re-discovery and Description of Original Material on the Photographic Researches of Sir John F. W. Herschel, 1839–1844," *The Journal of Photographic Science*, vol. 13, 1965, pp. 57–68.

Index

About the Author

GÜNTHER BUTTMANN is a Herschel scholar whose book *John Herschel* was first published in Germany in 1965 and was internationally hailed as a major biographical success. His earlier book, *Wilhelm Herschel*, was published in Germany in 1961. He was born and educated in Munich and now resides in Stockdorf, Germany, with his wife and three sons.